Introduction to Energy Technologies for Efficient Power Generation

Nature seeks and finds transition from less to more probable states

Ludwig Botzmann

The book was written based on a series of lectures as part of an academic course on "Heat technics and Power engineering." It is structured as to cover the horarium of 45 hours of a teaching plan and it comprises six modules

- Theoretical basis of energy conversions and thermodynamics
- Thermal motors (thermal engines)
- Thermal-power energy technologies
- Regeneration and recuperation of waste energy
- Energy transfer and accumulation
- Energy and indoor conditions

Alexander V. Dimitrov

Introduction to Energy Technologies for Efficient Power Generation

Alexander V. Dimitrov

CRC Press
Taylor & Francis Group
Boca Raton London New York

CRC Press is an imprint of the
Taylor & Francis Group, an **informa** business

CRC Press
Taylor & Francis Group
6000 Broken Sound Parkway NW, Suite 300
Boca Raton, FL 33487-2742

First issued in paperback 2020

© 2017 by Taylor & Francis Group, LLC
CRC Press is an imprint of Taylor & Francis Group, an Informa business

No claim to original U.S. Government works

ISBN 13: 978-0-367-57385-0 (pbk)
ISBN 13: 978-1-4987-9644-6 (hbk)

Library of Congress Cataloging-in-Publication Data

Names: Dimitrov, Aleksandŭr Vasilev, author.
Title: Introduction to energy technologies for efficient power generation/
    Alexander V. Dimitrov.
Description: Boca Raton : Taylor & Francis, a CRC title, part of the Taylor &
    Francis imprint, a member of the Taylor & Francis Group, the academic
    division of T&F Informa, plc, [2017] | Includes bibliographical references
    and index.
Identifiers: LCCN 2016038520| ISBN 9781498796446 (hardback : alk. paper) |
    ISBN 9781498796453 (ebook)
Subjects: LCSH: Electric power production--Energy conservation. | Electric
    utilities--Energy conservation. | Energy consumption.
Classification: LCC TK1005 .D56 2017 | DDC 621.31/21--dc23
LC record available at https://lccn.loc.gov/2016038520

Visit the Taylor & Francis Web site at
http://www.taylorandfrancis.com

and the CRC Press Web site at
http://www.crcpress.com

Contents

List of Figures

List of Tables

Acknowledgments

I thank Distinguished Professor Robert, PhD, Eng (UNLV) and Associate Professor Robert Kazangiev, PhD, Eng (IM-BAN) for their great help in this English edition of the book.

Finally, I thank my wife Zita Dimitrov, 1st degree research fellow (Eng) for rendering technical help and her patience during my writing activities.

Author

Alexander V. Dimitrov is a professional lecturer with 35 years of experience at four different universities. In addition to universities in Bulgaria, Dr. Dimitrov has lectured and studied at leading scientific laboratories and institutes in other countries, including the Institute of Mass and Heat exchange "Likijov," Byelorussian Academy of Sciences, Minsk; Lawrence Berkeley National Laboratory, Environmental Energy Technology Division, Indoor Environment Department; UNLV College of Engineering, Center for Energy Research, Stanford University, California, Mechanical Department.

Professor Dimitrov has conducted systematic research in energy efficiency, computer simulations of energy consumption in buildings, the distribution of air flow in an occupied space, modeling of heat transfer in the building envelope, and leaks in the ducts of HVAC (heating, ventilation, and air conditioning) systems.

Professor Dimitrov has defended two scientific degrees: Doctor of Philosophy (PhD) (in 1980) and Doctor of Science (DSc) (in 2012). With significant audit experience in the energy systems of buildings and their subsystems, he has developed an original method for evaluating the performance of the building envelope and energy labeling of buildings. He also has experience in the assessment of energy transfer through the building envelope and ducts of HVAC systems. Professor Dimitrov's methodology has been applied in several projects with great success. He has developed a mathematical model for assessment of the environmental sustainability of buildings, named BG_LEED. He earned his professor degree in "engineering installations in the buildings" with the dissertation "The building energy systems in the conditions of environmental sustainability" at the European Polytechnical University in 2012. Subsequently, he has authored more than 100 scientific articles and 9 books, including 4 in English.

Abbreviations

AC	Air conditioning
ACS	Automatic control system
BACS&M	Building Automatic control system and monitoring
CS	Cold source
CW	Construction works
DDC	Direct digital control
DTEG	Direct thermo electrical technology
EC	Energy center
ED	Energy department
EE	Energy efficiency
FC	Fuel cell
FES	Fossil energy sources
GBA	Gross building area
GDP	Gross domestic product
HDT	Hydro dynamical technology
HI	Heating installation
HP/HV	High pressure/voltage
HS	Heat source
ICE	Internal combustion engine
InDET	Indirect thermo electrical technology
LDC	Lower dead center
LH/LV	Low pressure/voltage
LtDC	Left dead center
MHG	Magneto hydrodynamical generator
NPS	Nuclear power station
NSTA	Nuclear steam turbine assembly
PhEA	Phase exchange accumulation
PSHS	Pumped-storage hydropower station
PV	Photovoltaic cell
RDC	Right dead center
RES	Renewable energy sources
TDC	Top dead center
TDP	Thermodynamical process
TDS	Thermodynamical system
TPS	Thermal power station
TWC	Typical weather conditions

Introduction

The utilization of energy is the key to understanding modern civilization. The field of energy science began its successful development after the 18th century, at which time a majority of important scientific discoveries related to energy occurred. It was discovered that energy could cause global ecological catastrophes and even the destruction of the human race, if not properly utilized.

The book *Introduction to Energy Technologies for Efficient Power Generation* reveals the technologies for the production, storage, and transfer of energy related to the energy and building industries, and transportation. It is intended to be a reference tool for current and future leaders, managers, and specialists in these areas, which will help them with the effective and environmentally friendly utilization of energy resources.

Together with the classic energy technologies for energy conversion, discovered by W. Rankine, G. Brayton, R. Stirling, and N. Otto, the book describes some of the technologies less known by the general public, such as

- The osmotic technology of D. Bernoulli
- The thermoelectric technology of T. Seebeck
- The photovoltaic technology of A. Becquerel/W. Shottky
- The technology of the controlled oxidation of fuels of C. Schonbein/W. Grove in the fuel cells
- The vortex technologies of G. Ranque and Merkulov/Y. Potamov
- The cold nuclear fusion technology of A. Rossi

The book also covers the technologies of regeneration and recuperation of waste industrial energy, as well as technologies for the generation of energy from renewable resources, such as solar, wind, and geothermal.

The book describes the contemporary concepts for thermal and electric energy transfer through solid bodies at the microscopic thermodynamical level. Also discussed are the methods for effective control of energy fluxes in technical equipment and appliances, including building envelopes.

Finally, also discussed is the work organization as well as desired competencies of energy managers for industrial, public, commercial, and residential buildings. The book focuses on the role of organizational leaders for the maintenance and proper usage of the systems responsible for the internal comfort of the environment, that is, temperature, lighting, and clean air in the serviced buildings. Also included is a description of the systems for thermal comfort and their elements.

Chapter 1: Theoretical Foundations of Energy Conversions and Thermodynamics

Chapter 1 treats basic notions and laws of energy conversion—the principles of thermodynamics. For instance, "ideal gas" and Einstein "ideal solid" are presented as working bodies where energy conversion takes place. Laws of energy conversions and their directions are formulated in detail, and the types of thermodynamic process are also explained. The basic requirements to power machines (engines, generators/converters, etc.) are set forth and machine classification from a functional point of view is given. Energy efficiency is also discussed employing the ideal thermal engine of Nicolas Leonard Sadi Carnot as an energy standard.

In addition, the first part clarifies the characteristics of various types of real energy conversion. For instance, details concerning a working body where phase conversions take place are outlined, or compressibility of working fluids and interatomic forces, Ranque vortex effects, and Joule–Thompson inverse temperatures and their applications in the heat generators of Potapov and Hirsch are accounted for.

The next chapters of the book discuss some basic types of energy conversion referred to as thermal-energy technologies, such as

- Heat-mechanical energy conversion
- Heat-electric energy conversion
- Regeneration and recuperation of thermal energy

Chapter 2: Conversion of Thermal Energy into Mechanical Work (Thermal Engines)

The development of thermal engines and steam and gas turbines, in particular, yielded significant change in power technologies, unleashing mass exploitation of electricity in households, industry, and transport.

Initially, the so called externally fired thermal engines were designed. These were the steam engines of Denis Pippin, Thomas Newcomen, and James Watt, the engines of Stirling, the steam turbine of Parsons, and the gas turbine of John Barber. They were known for the heat generation devices (combustion chambers and boilers) that were physically separated from the operating mechanisms (power cylinders and mechano-kinematical chains). Moreover, accessible and cheap heat sources were used—wood biomass and its derivatives (firewood was mostly used till the end of the 19th century), charcoal and coal (their use started at the beginning of the 15th century).

An improvement of thermal engines gathered momentum in the mid-19th century and consisted of efficiency increase and the minimization of the overall dimensions of engines. The innovative idea was to combine the combustion chamber and the power cylinder into one solid machine unit. The "internal combustion engine" materialized this new concept. Otto and Langen were the first to practically realize the idea in 1858. Later, Atkinson (1882), Rudolf Diesel (1890), and Miller (1947) modified and improved it, while a modern version of the engine is that of Felix Wankel (1951). The Brayton engine and its modification in the gas-turbine and gas-compressor engines are also described. Thermomechanical technology employing external heat sources are presented by the Stirling engine.

Each thermal technology discussed herein will be illustrated by specific physical schemes and devices.

Chapter 3: Thermoelectric and Co-Generation Technologies

This chapter discusses some energy-conversion technologies where heat is directly converted into electricity or is a previously created product by energy conversions. We shall analyze the following technologies:

- Conversion of solar radiation into electricity (the so called "internal ionization")
- Seebeck technology applied in thermos-couples (the so called "thermoelectric current")
- Schonbein/Grove technology with oxidation control
- Rankine cycle technology
- Brayton cycle technology

The advantages of the co-generation processes and devices operating within the Brayton clockwise cycle and their adaptively to household energetics are underlined. The employment of thermoelectric and cogeneration technologies in the design of vehicle hybrid gears is also discussed.

Each thermal technology discussed herein will be illustrated by specific physical schemes and devices.

Chapter 4: Energy Rehabilitation: Regeneration and Recuperation

In this chapter, the technologies for energy rehabilitation are classified in two groups as shown in

- Regeneration technologies, where the energy potential of the working fluid is used directly or increased by keeping the energy form—generally by increasing the temperature or pressure of the authentic energy-charged medium
- Recuperation technologies, where the available energy potential of the working fluid is used to change the energy form (for instance, "heat or mechanical motion to electricity" at the expense of a more complete use of the residual enthalpy)

Some new modern technologies of electric power generation are disclosed—osmotic technology, hydro dynamical technologies, solar concentrators, and that of low-temperature plasma, magnetic hydrodynamic generators.

Each thermal technology discussed herein will be illustrated by specific physical schemes and devices.

Chapter 5: Energy Transfer and Storage

Technologies of energy transfer and storage are also addressed. Four different microscopic level models are based on hypothetical mechanisms that were developed during the 1940s. These are

Kaganov's model or a two-step mechanism of interaction between phonons and electrons, Geier's model and Kumhasi's model, A. Majumdar's model and model of lagging temperature gradient.

In the bodies under the influence of the environment (external energy fluxes or electromagnetic or gravitational fields) directed movement of micro particles occurs

- Type fermions (electrons and ions), or
- Quasi particles (bosons type: photons or phonons by means of internal ionization or polarization of the atomic structures)

The movement of these particles in the transmission medium is hampered by the body's structural components (atomic lattices or molecular entities). This process is accompanied by transfer of an amount of movement from the quasi particles (bosons) flow toward the lattice, whose internal energy (respectively temperature) is changed, proportional to the flux value. Thus, the change in the body's temperature is caused by the directed movement of quasi particles, delaying (lagging) in time relative to the passage of their flow. The four models of heat transfer in solids contribute to the understanding of the physical character of transfer.

We shall discuss in what follows their features and practical methods of their regulation by the introduction or avoidance of thermal resistance, such as volume or Fourier resistance, capacitive resistance (due to storage), fluid dynamic resistance (due to convection), and reflection resistance (due to radiation).

When the working body of a thermodynamic system is solid, the transferred energy is assessed by the total energy transfer law, giving the relation between the specific energy flux **h** and the gradient of free energy potential in the solid body.

Chapter 6: Energy and in Door Enviroment

Specific environmentally friendly energy conversions in inhabited areas of buildings and passenger vehicles are outlined at the end of the chapter.

A favorable and comfortable indoor environment can be attained employing two architectural-engineering technologies which complement each other:

- Passive technologies
- Active technologies

The first type exploits the functional properties of the envelope of an object (building or vehicle) to filter environmental impacts such as solar radiation, wind, humidity, and aerosol emissions. Once designed and laid, the envelope operates as an insulation saving energy needed to provide a friendly indoor environment.

The second type uses energy generated in central or local power stations. That energy is needed by the devices to change the parameters of the indoor environment, offering thermal, illumination, and air comfort.

The systems of thermal comfort regulation are divided into three basic groups with respect to the mechanisms providing comfort in the inhabited area:

- *Convective systems.* The necessary local conditions for thermal comfort in the inhabited area are created by convective heat transfer and change of the air temperature T_a
- *Radiative systems.* Comfort via those systems is attained mainly by change of the radiant temperature T_R of the heaters located around the inhabited area. Air temperature T_a is generally kept constant
- *Convective–radiative systems.* Temperature T_a as well as temperature T_R vary in the conditioned area

Considering the working fluid in the distribution network, the systems are subdivided into: air-to-air; water-to-air; steam-to-air; electricity-to-air; and gas-to-air. With respect to heat transfer, they are classified as

- Systems using natural convection
- Systems with forced convection

Systems providing thermal comfort (air conditioning or heating) reflect the specificity of the objects involved. Hence, engineering offers a number of systems which differ from each other with respect to the internal arrangement of heat distribution. There are four basic methods of air cleaning (purification) and control of the contaminants in rooms and passenger compartments (applied independently or combined with each other): source elimination or modification; use of outside air to remove contaminants (space ventilation); local ventilation; and air cleaning. Each thermal and air cleaning technology discussed herein will be illustrated by specific physical schemes and devices.

The last part of the Chapter 6 discusses the work organization as well as desired competencies of energy managers for industrial, public, commercial, and residential buildings. The book focuses on the role of organizational leaders for the maintenance and proper usage of systems responsible for the internal comfort of the environment, i.e., temperature, lighting, and clean air in the serviced buildings.

1

Theoretical Foundations of Energy Conversions and Thermodynamics

1.1 Energy, Energy Conversion, and Generalized Work

1.1.1 Introduction

This book is intended to be introductory for all who attend academic courses on "Organization and management of industrial and transport companies (industrial and transport management)," students at the "electrical" faculties of technical universities (e.g., faculties of energetics, electric power stations, and networks and systems), and those who are "seduced" by the challenges of energy efficiency including energy supervisors of industrial installations and structures (architects, designers, contractors, and investors). The book also addresses curious readers of engineering-technical literature, persons who are directly involved in investments in new energy-related (power) technologies such as contractors, civil servants, bank project supervisors, or simply people who see the topic of energy as an intellectual challenge.

This book introduces basic terms and laws concerning energy conversion from one form into another, yet from an engineering point of view and with a particular focus on transport and industrial applications. As is known from physics, more than 20 processes of energy conversion operate (see Table A.1 in the Appendices), but only four of them have been "*commercially*" recognized. These are

- Oxidation of chemical substances and compounds (including "controlled oxidation" as in the process of Schonbein/Grove)
- Electromagnetic induction
- Internal ionization (photovoltaic effect)
- Nuclear fission (by splitting large atoms into small)*

Their common characteristic is in that some type of work is done (chemical, mechanical, electrical, or nuclear), conversion of one type of energy into another type of useful energy takes place, and heat is released. Note that heat is an inevitable (desired or not) accompanying product of all energy conversions. Hence, it is worth knowing all details of its generation, conversion, transportation, and usage.

The first part of the book treats basic notions and laws of energy conversion—the principles of thermodynamics. For instance, "ideal gas" and Einstein "ideal solid" are presented as working bodies where energy conversion takes place. Laws of energy conversions and their directions are formulated in detail, and the types of thermodynamic process (TDP)

---

* Nuclear fusion (creating large atoms by a merger of smaller atoms) is many years away from commercial use.

are also explained. The basic requirements to power machines (engines, generators/converters, etc.) are set forth and machine classification from a functional point of view is given. Energy efficiency is also discussed employing the ideal thermal engine of Nicolas Léonard Sadi Carnot as a standard.

In addition, the first part clarifies the characteristics of various types of real energy conversion. For instance, details concerning a working body where phase conversions take place are outlined, or compressibility of working fluids and interatomic forces, Ranque vortex effects, Joule–Thompson inverse temperatures, etc. are accounted for.

The second part of the book discusses some basic types of energy conversion referred to as thermal-energy technologies, such as

- Heat-mechanical energy conversion
- Heat-electric energy conversion
- Regeneration and recuperation of thermal energy

Technologies of energy transfer and storage are also addressed. Specific environmentally friendly energy conversions in inhabited areas of buildings and vehicle passenger vehicles are outlined at the end of the chapter. Some new modern technologies of electric power generation are disclosed—osmotic technology, Seebeck's technology, technology of solar concentrators and that of low-temperature plasma, magnetic-hydrodynamic generators (MHGs), technologies employing classical fuel cells and photovoltaic solar panels, etc. The advantages of the cogeneration processes and devices operating within the Brayton clockwise cycle and their adaptively to household energetics are underlined.

The third part discusses available energy resources and energy strategies employed to decrease the import of energy sources by increasing the efficiency of energy generation and reducing losses via energy saving in basic economy trends—transport, industry, and households. Last, some problems of the efficient use of primary energy resources and environmental protection are also given.

1.1.2 Historical Survey of the Stages of Energy Utilization

From a gnoseological point of view, a survey of knowledge and experience in energy utilization seems useful. The evolution of civilization depends on the accumulation of human knowledge and practical experience in energy mastering. Hence, the stages of the latter correspond to those of the history of mankind. In this respect, several periods are outlined:

1. Primitive (savage) period
2. Initial period (corresponding to slave-holding and early feudalism)
3. Renaissance and preindustrial period
4. Industrial period
5. Global era

Due to the small human population, the first period was characterized by scarce energy consumption using biomass as an energy source. People relied basically on natural energy sources (sunlight and geothermal energy) inhabiting areas with favorable climate, only. Those were areas in Mesopotamia (on the banks of Tigris and Euphrates) and on the banks of the Nile. Middle and Northern Europe and Asia were colonized right after humans learned how to survive outside caves and in severe climate. Hence, emerging though primitive thermotechnics may be assumed as one of the preconditions of the human migration to and colonization of new geographic areas.

In addition to the severe northern climate, humans faced the problem of covering vast areas. Trade and supply of raw materials needed for emerging crafts and the division of labor resulted in the manufacture of primitive vehicles—animal powered carts, boats, and ships. Historical proof of the use of natural energy sources in transport are scarce but ancient manuscripts claim that sailboats were used as far back as as the 12th to 13th century BC.

Naturally, society used human and animal power. In the slave-holding societies of Mesopotamia, along the Nile, and in the Mediterranean region, human power was treated as an article of trade, but the methods of their acquisition was violent—assaults, robberies, wars, and enslavement. These were the preconditions for numerous ancient campaigns; sea travels, etc., such as the Trojan War, the marches of the Persian kings on ancient Greece, marches of Alexander the Great, marches of Hannibal on Italy, etc. These campaigns were driven by the necessity of acquiring biological engines and by the need for biological energy. The more slaves an ancient ruler possessed, the richer and stronger he was and the larger were the territories he controlled. Slaves were exploited everywhere—in mining precious metals, copper, and iron; in workshops (to fabricate arms, dishes, and metal hardware); in farms and households (as servants, baby-sitters, and teachers); in shows (as gladiators, actors, and musicians); as bath attendants, etc.

The whole public production was based on slave labor. Specific markets existed for slave acquisition, distribution, and supply, and the climax of that "industry" was attained in the Roman Empire whose ruins were not discovered until the Middle Ages.

The human "engine," however, possessed two major disadvantages:

- It needed regeneration
- It needed motivation for work

It seems that the second disadvantage turned to be "the thorn in the flesh" of slavery. Slaves revolted and escaped, and masters realized their unreliability as a labor source. The increase and improvement of food quantity and quality, provided by the slave-owning means of production and being directly related to the slaves' exploitation, resulted not only in the increase of human population but also in material deficiency. This was so, since the need for worldly goods was greater than the capacity for their production or delivery, and delivery was mostly possible via military campaigns. Thus, the forms of production had to be modified, and feudal production emerged as a result. In the early Middle Ages, slaves were formally released and their biological energy was harnessed in the cultivation of the master's land, while they were chained to the land. Yet, serfs had more interest in labor than slaves. The growth of the human population proceeded to overrun productivity, and people found a motive to increase productivity by the division of labor and particular specializations. Hence, crafts and liberal professions emerged, giving birth to research in astronomy, mathematics, geography, etc.

Moreover, human population growth stimulated the human settling of new territories with severe climate, bringing forth the need for energy for heating. Wood mass was used as a heating source, and the English and German forests were cut down—and the French woods were destroyed in 12th to 13th centuries in the search of timber for cathedral construction. When this energy source was significantly reduced, coal was found to be a proper commercial substitute, easily transported, and with capacity larger than that of the wood biomass. Thus, coal was more and more intensively used at that time to satisfy household, craftsmanship, and industrial needs. This marked the beginning of the pre-renaissance era and the third stage of energy exploitation, while the period experienced a wider use of wind energy and the emergence of new thermal technologies of coal burning. Wind energy was harnessed mainly to mechanize some industrial operations, such as wheat milling, for instance. The sea-going fleet was entirely renovated—old row boats with slaves-rowers were replaced by sailing ships (frigates).

Better ships and wider knowledge were the basic prerequisites for the emergence of a new era in human civilization—that of great geographic discoveries. However, a race (almost all military based) for colonization and redistribution of the newly discovered territories, and exploitation of their natural resources started between the European nations. After the conquer of the New World with its human and material resources, an industrial period of European civilization started, initially in England in the 18th century and later in other countries and on other continents.

The earliest designs of energy-fed devices may be referred to this period—such were the steam boiler and steam engine of Papin* (1678r.), for instance. A real thermal steam engine was constructed by Thomas Newcomen in 1712, it was later significantly improved by James Watt (1765) and used in mining. However, the technical revolution did not bypass transport. The first automobile with a steam engine was constructed in 1769 by Nicolas Cugnot, and 32 years later (1801) Richard Trevithick designed the first railway steam locomotive in London.

The foundations of modern energetics were laid in the 19th century which may be referred to as the golden century of energy. Here is the chronology of some key events:

- Discovery of the electromagnetic induction effect and invention of the first electrical generator (1826–1832)
- Discovery of controlled oxidation and design of the first fuel cell (1838–1842)
- Discovery of internal ionization and the photovoltaic process (1839)
- Construction of the first stationary internal combustion engine (1858)
- Siemens dynamo marking the initiation of worldwide electrification (1867)
- George Brayton constructed the reciprocating oil-burning engine with internal combustion (1871)
- Gustav de Laval (1883r.) developed the first active turbine and Charles Parsons (1884)—the first reactive steam turbine
- Nikola Tesla constructed the air conditioner (AC) generator (1888)

1889 was the year when the design of cars for common use started in France. Thus, the time of mass car transport† came, unleashing the frantic need for new energy sources for engines and marking the beginning of the era of oil‡ and its derivatives that continue to this day.

---

* The first closed kitchen stove was designed by Francois Cuvillies, an architect, in 1735.
† The *Model T* of Henry Ford marked the beginning of car mass transport. It was manufactured applying a new conveyor-belt technology within the period 1908–1922 (time of car assembly was 1.5 h). A two-cylinder engine was incorporated in the vehicle, and fuel consumption was 0.05 gal/mile. The car price was $800 and affordable for the majority of people. The Model T gathered a velocity of 21 miles/h conforming to the state of roads at that time.
‡ W. Churchill as First Lord of the Admiralty in 1911 made a historical decision to change the power of warships, and old steam engines were replaced by internal combustion engines.

In 1970–1972, during the first great OPEC oil embargo, government officials found that the strong need for cartel-traded energy sources resulted in energy dependency by the former colonies. That dependency has not been overcome even now since about 50% of the EU energy sources are imported. This book describes innovative technologies for energy generation. Applied on a sufficiently wide scale, these technologies can guarantee to a large extent the energy independence of a particular state and its people.

1.1.3 Basic Terms of Energy Conversion and Thermodynamics

1.1.3.1 Energy

Energy is a basic physical and engineering term originating from the Greek word "ενεργεία ≡ activity." It is the property of material systems to interact, and interaction takes place between material bodies, energy fields, or between bodies and fields (electric, magnetic, gravitational, etc. fields are known as energy fields). As noted, work is done and heat is released during these interactions. The following interactions in Table 1.1 are known as corresponding to the energy forms.

The work done can be "visualized" via the ordered motion of material particles,[*] which is accompanied by imposed, directed motion of some of them.[†] As for classical physics, chaotic motion is considered to be "energy waste" or "poor" (*low quality*) heat. Work is treated as directed, useful, and organized form of motion while heat is considered to be coproduced by aimless movable quasi-particles due to particle interaction (Dimitrov A.V., 2015). During energy-based interactions, all system parameters can vary. These are

- Available quantity of matter, electrical quantity, volume occupied by matter, and chaos of motion
- Temperature, pressure, electric potential, and internal energy

The first group of parameters is called *generalized displacements* (or *coordinates*) while the second one—*generalized forces* or *potentials*.

In brief, energy exchange between bodies and energy fields is measured introducing "energy quantity." The SI measure used is Joule (1 [J] = 1 Nm) and its derivatives

TABLE 1.1

Energy Forms and Their Corresponding Works

Interactions		Energy Forms
Mechanical interaction	⇒	Mechanical energy
Thermal interaction	⇒	Thermal energy
Electromagnetic interaction	⇒	Electrical energy
Chemical interaction	⇒	Chemical energy
Nuclear interaction	⇒	Nuclear energy

---

[*] Electrons, ions, photons, phonons, and others (see in Chapter 5 for details).
[†] The vector quantities of energy transfer (motion of the energy carriers) are described through the total transfer low (see in Chapter 5).

(1 kJ = 10^3 J, 1 MJ = 10^6 J, 1 GJ = 10^9 J, and 1 TJ = 10^{12} J).* There are also other systems of energy measurement such as the IP system used in the Anglo-Saxon world or the antiquated but still popular system "kg–sm–h." The link between different measurement systems is given in Table A.2—see the Appendices.

1.1.3.2 Thermodynamic System

According to modern science, a thermodynamic system (TDS) is a set of bodies and energy fields exchanging energy and mass between each other. It possesses expressed geometrical boundaries called *"control surfaces"* while its interior is treated as a *"control volume."* All material objects outside the control surface constitute TDS surroundings (environment).

Considering links with the environment, the TDSs may be classified in three groups:

- *Insulated TDS* (without exchange of mass and energy). A special case in this group is the adiabatic TDS (not exchanging energy).
- *Closed TDS* (no exchanging mass).
- *Open TDS* (exchanging energy and mass).

There are two TDS types with respect to state: nonequilibrium and equilibrium. If the physical parameters of a certain TDS are kept stable under specific conditions, then the TDS is in equilibrium.

Violation of TDS equilibrium may take place if the system undergoes mechanical, thermal, chemical, electromechanical, or nuclear impacts. The number of nonequilibrium states determines the TDS degrees of freedom.

Classical thermodynamics considers TDS with two degrees of freedom, only, including mainly *mechanical and thermal nonequilibrium.* Those nonequilibrium states are expressed as *expansion/compression* or *heating/cooling* of the control volume. Modern needs, however, imply the study of more complex TDS—for instance, those with three degrees of freedom where three types of nonequilibrium are observed (thermal, electrical, and chemical). Actual examples of such TDS are fuel cells, electric accumulators, absorption heat pump engines, etc.

To establish the equilibrium state of a TDS, one needs to study all TDS characteristics called *"state parameters."* The latter are divided into two groups: *basic and calorimetric.* Temperature T, volume V, pressure p, and the electric potential are *basic state parameters,* since they can be directly measured using analog or digital instruments. Internal energy U, entropy S, enthalpy H, and free energy potential D cannot be measured but can be calculated, only, after performing appropriate modeling of the processes that take place in a TDS. Hence, these parameters are known as *calorimetric ones.*

1.1.3.3 Basic State Parameters

The main state parameter in a TDS with two degrees of freedom is temperature T. It is assumed that temperature is an indicator of the quantity of internal energy U available in the neighborhood of point A of a TDS or within its control volume. Four worldwide

---

* Popular literature uses the quantity of a primary energy carrier as a measure of energy consumption. This quantity is expressed in metric kilograms of oil equivalent— kg oe or toe (1 toe ≡ 10^3 kg oe). Relation 1 toe ≡ 42 GJ can be used to express conversion to SI units. Table A.5 in the Appendices specifies the coefficients of conversion of the measurement units of various energy carriers used in practice.

TABLE 1.2

Ref Marks of the Used Temperature Scales

	Zero Point	**Step**	**Melting Point—T_m**	**Boiling Point—T_b**
Celsius	0.0	100.0	0.0	100.0
Kelvin	−273.16	100.0	273.16	373.16
Fahrenheit	−32.0	180.0	32.0	212.0
Rankin	0.0	180.0	0.0	180.0

approved temperature scales exist: Celsius (°C), Kelvin (K), Fahrenheit (°F), and Rankine (R). Their common feature is the use of common bench marks—the two characteristic temperatures of water (the most widely spread compound on Earth): temperature of water solid–liquid transition (melting point) and temperature of water liquid–vapor phase transition at pressure of 1.10^5 Pa (boiling point). The four scales differ from each other with respect to the choice of a zero point and grading of the temperature range between the bench mark temperatures (see Table 1.2).

The link between different temperature scales is given by the following relations:

$$T(^\circ C) = \frac{5[T(F) - 32]}{9}, \quad T(F) = 1.8 \cdot T(K) + 32,$$

$$T(R) = 1.8 \cdot T(K) \quad \text{and} \quad T(K) = t(^\circ C) + 273.16.$$

Here, T(°C), T(R), T(F), T(K) are temperatures measured using the scales of Celsius, Kelvin, Fahrenheit, and Rankine.

The second basic state parameter is pressure p, since it contains information on the potential energy available in the neighborhood of a certain point or within the volume of the control area (see Figure 1.1). Pressure unit is Pascal [Pa], where 1 Pa = 1 N/m² which is a very small quantity. For instance, the value of the atmospheric pressure at sea level is 101,325 Pa or 0.101 MPa.

Common pressure values are much larger than the value of the atmospheric pressure, and the use of the multiplier 10^5 complicates the speech communication between specialists. Hence, 6 ÷ 7 more units have become popular in engineering such as Bar, at, mm_{H_2O}, mm_{Hg}, psi (pound per square inch), and psf (pound per square foot).[*]

Other TDS state parameters of interest are

- Mass of the working body (m_{TDS}, kg; K_{mol}; M_{kg})
- Volume of the control area V_{TDS} (m³)

and complexes:

- Specific volume [$v = V_{TDS}/m_{TDS}$, m³/kg]
- Density [$\rho = v^{-1} = m_{TDS}/V_{TDS}$, kg/m³]

The performance of theoretical calculations and the preparation of engineering projects involve the use of so-called "calorimetric" state parameters—internal energy U, entropy

---

[*] Fluids physical properties are determined via measurement of the absolute pressure in the following units: 1 Bar = 1 at = $10 m_{H_2O}$ = $10^5 mm_{H_2O}$ = 736 mm_{Hg} = 14.69 lbf/in² = 2115.10^3 lbf/ft² = 30 in_{Hg} (absolute pressure is the sum of atmospheric pressure and overpressure).

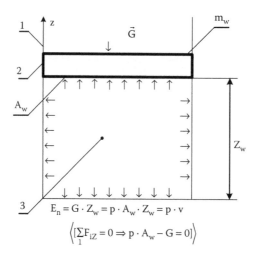

FIGURE 1.1
Potential energy for v = const will depend on the pressure values, only: 1—cylindrical control surface; 2—moving wall-control surface; and 3—working fluid (body) of the TDS. Here: G—reduced force acting on the moving wall and E_n—potential energy.

S, enthalpy H, and free energy potential D. Commonly, all of them are important from a physical point of view, since the change of the internal energy U is an indicator of energy storage in the working body or TDS. Enthalpy H* indicates the body's entire energy available, since it is a sum of the internal kinetic and potential energies—(U) and (pV), that is, H = U + pV. Entropy S in its turn is a measure of energy organization in a TDS, indicating the system degeneration (see in Dimitrov A.V. 2015).

   Last but not least in the list of TDS calorimetric characteristics is the free energy potential Ψ.[†] It is a universal engineering characteristic used in noninsulated and open TDS, since it contains information about the energy charge of the control volume and energy conversions expected to take place within this volume. Additional information on the calorimetric units can be found in Sections 1.1.5 and 1.1.6.

1.1.4 General Information on the TDS Working Body (Medium)

This introductory part of thermotechnics, similar to other general engineering disciplines such as mechanics, strength of materials, and fluid mechanics, operate with *hypotheses and idealized physical models*. As far as mechanics employs models of "mass point" and "absolute solid," strength of materials treats absolute elastic bodies, and fluid mechanics deals with potentially vortex-free flows, the following models are introduced in thermodynamics:

- Ideal gas (fluid) is a working body where TDPs run but forces acting between atoms and molecules (the so-called van der Waals forces) and atom and molecule volumes are *neglected*, that is, dimensions of atoms and molecules are disregarded, and they are treated as geometric points. Yet, friction forces in an open TDS are exclusively due to viscosity, and vortex effects are neglected.

---

* Introduced by J.W. Gibbs (1839–1903) in 1875r. For a unit mass, the enthalpy (ἔνθαλπος) is h = u + p · v.
† Introduced by P. Duhem (1861–1916), see Dimitrov A.V. (2015).

- Einstein ideal solid body is a working body composed of discrete atoms and molecules where energy conversion and specific processes take place, but interactions between body components are neglected—each atom in the lattice is considered to be an independent three-dimensional (3D) quantum harmonic oscillator, and all atoms oscillate with the same frequency. In what follows we shall discuss some of the basic properties of an idealized working body as a TDS component.

1.1.4.1 Ideal Gas

Classical *kinetic gas-molecular theory* proves that gas pressure p in a unit control volume can be calculated via

$$p = \frac{2}{3} n_A \cdot E_k = \frac{2}{3} n_A \cdot \frac{m \cdot \overline{w}^2}{2}, \tag{1.1}$$

where

$n_A = N_{Env}/V_{Env}$ is the number of molecules per unit volume (and N_{Env}—number of molecules in the control volume).

$E_k = m.\overline{w}^2/2$—kinetic energy of gas molecules (m—mass and $\overline{w}^2$—squared mean velocity of molecules).

The following expression of pressure is valid for the whole control volume V_{Env}:

$$p = \frac{2}{3} aT \frac{N_{Env}}{V_{Env}}. \tag{1.2}$$

It is assumed that the kinetic energy E_K is proportional to temperature T of the ideal gas, and the substitution $E_k = a \cdot T$ is thus used. Then, Equation 1.2 takes the form[*]

$$\frac{pV}{T} = \frac{2}{3} a \cdot N_{Env} = const. \tag{1.3}$$

Reduced to 1 mass (unit mass), it reads

$$\frac{pv}{T} = \frac{2}{3} a \cdot N_A. \tag{1.4}$$

The mathematical form found proves that TDS basic parameters (p, T, and v) form a complex quantity at equilibrium. It remains constant regardless of the change of parameter absolute values upon TDS occupying a new state equilibrium. Equation 1.4 was derived by Clausius and Clapeyron[†] and named after them. It predicts the values of the characteristic parameters of a postequilibrium TDS.

---

[*] Under standard normal conditions (p_n = 101,325 Pa and T_n = 273,16K): N_A = 8314 J/kmolK, N_A—number of molecules per unit mass (1 kμol).
[†] R. Clausius (1822–1888) and B. Clapeyron (1799–1864).

Assume that a certain TDS is characterized by pressure (p_1), temperature (T_1), and specific volume (v_1) at a certain moment of time. Under external impacts (compression, expansion, heating, or cooling), pressure (p_2), specific volume (v_2), and temperature (T_2) acquire new values, and the TDS occupies a new equilibrium state. According to Equation 1.4, the following relation between the two equilibrium states holds

$$\frac{p_1 v_1}{T_1} = \frac{p_2 v_2}{T_2} = \text{const.}$$

The new equilibrium parameters can be calculated via the expressions

$$v_2 = v_1 \frac{p_1}{p_2} \frac{T_2}{T_1} \quad \text{or} \quad T_2 = T_1 \frac{p_2}{p_1} \frac{V_2}{V_1}.$$

It follows from the Clausius–Clapeyron equation that all gases subjected to identical pressure (p) and with identical temperature (T) of the control volume (V) have identical number of molecules $N_A = 6022.10^{26}$ num/mol = const. This fact was established by Avogadro[*] and is known as Avogadro's number.

Under normal conditions (for p_n and T_n), an ideal gas with mass 1 кmol[†] will occupy a control volume of 22.4 m³. This fact follows from Avogadro's law and is of crucial importance in finding other working parameters of an ideal gas at normal conditions, namely:

- Density (ρ_n)—$\rho_n = M/22.4$ kg/m³
- Specific volume (v_n)—$v_n = 22.4/M$, m³/kg
- Gas constant (R_μ)—$R_\mu = R/M = 8314/M$, J/kµolK

while Clausius–Clapeyron law will take the form

$$p \cdot v = R_\mu \cdot T. \tag{1.5}$$

For different gases, constant R_μ has different values (see Table A.3 in the Appendices).

Another important property of the ideal gas is that the energy accumulated can be assessed by means of the derivative of gas internal energy (U) with respect to temperature (T):

$$\text{Accumulated energy} \approx \frac{dU}{dT}$$

Using the equations of the kinetic gas-molecular theory of heat written per unit mass, the above expression takes the form

$$\frac{\text{accumulated}}{M} = \frac{du}{dT} = \frac{3N_A}{M} \approx 3R_\mu = \text{const} \tag{1.6}$$

(for $T \geq 100$ K). This formula was derived and proved by A. Petit and P. Dulong[‡] in 1819.

---

[*] Avogadro A. (1776–1856).

[†] 1—kilogram per mass of ideal gas is equal to gas molar weight $M = \Sigma_1^n A_i$, where A_i is Mendeleev's atomic weight. For instance, $M_{CO_2} = A_C + 2A_0 = 12 + 216 = 44.0$ kg; hence, 1 кmol of CO_2 contains 44 kg matter.

[‡] A. Petit (1791–1820) and P. Dulong (1785–1838).

Dulong–Petit law states that the ratio du/dT is constant and does not depend on body temperature. Since du/dT characterizes each fluid as a working fluid accumulating heat, it is called specific heat capacity. It is the measurable physical quantity that shows the amount of heat required to change by 1 K the temperature of 1 kg mass of a TDS working body. Heat capacity is denoted by C_v ($3R_\mu = C_v$) and is experimentally found for each chemical compound regarding a constant TDS volume. The values of C_v are specified in Table A.3 of the Appendices.

It follows from Equation 1.6 that the change of the internal energy du per unit mass is equal to

$$du = C_v dT. \tag{1.7}$$

Then, the final increase of the internal energy Δu in two successive states of an ideal gas at temperature change $T_1 \rightarrow T_2$ can be written in the form

$$\Delta u = C_v(T_2 - T_1), \tag{1.8}$$

and the final value of the internal energy u_2 will read

$$u_2 = u_1 + C_v(T_2 - T_1). \tag{1.9}$$

Besides its measurement at constant volume (v = const), the specific heat capacity can be measured or assessed at constant pressure (p = const), too. For ideal gases, there is a difference between the two values in favor of the second one, and it is expressed by Meyer's law

$$C_p - C_v = R_\mu. \tag{1.10}$$

The specific heat capacity C_p is a physical quantity illustrating capacity of a working fluid to accumulate entire energy (enthalpy h = u + pv) per unit mass

$$C_p = \frac{dh}{dT}.$$

Hence, enthalpy increase dh will read

$$dh = C_p \cdot dT \quad \text{or} \tag{1.11}$$

$$\Delta h = C_p(T_2 - T_1). \tag{1.12}$$

Enthalpy final value at temperature T_2 can be written as

$$h_2 = h_1 + C_p(T_2 - T_1). \tag{1.13}$$

The ideal gas model is used to predict the state of a TDS considering all processes of energy conversion where the working fluid keeps its vapor state. These are for instance processes running in heat engines with internal combustion, gas turbines, turbo-reactive and gas-turbine engines, air refrigerators, and other engineering devices where no phase transitions take place.

1.1.4.2 Einstein Ideal Solid

By the end of the 19th and at the beginning of the 20th century, *the kinetic gas-molecular theory* predominated in thermotechnics. It was based on the concept of material continuum and successfully treated working gases and solids. Engineering practice proved that solids, similar to gases, were a medium capable of accumulation and transfer of energy (including heat). It was deduced that due to the small volume expansion, solids as compared to gases had specific heat capacity C_v equal to C_p. Hence, the Dulong–Petit law was extrapolated for solids in the form

$$C_v = C_p = 6.4 \, cal/gm - mol/k = const, \tag{1.14}$$

showing that solid capacity to accumulate energy did not depend on temperature.

Enough evidence was accumulated by the end of the 19th century that the Dulong–Petit law did not hold true for low temperature (T < 100 K) and it was found that $C_v = f(T)$. Moreover, at temperatures close to the absolute zero, solids lost their capacity to accumulate energy and $C_V = C_P \approx 0$. This could not be explained by the classical heat theory that assumed solids to be a continuous medium, and a "heat crisis" started and gathered momentum in science. Albert Einstein in 1905 was the first to explain the phenomenological dependence of heat capacity of solids on temperature. Considering the Max Planck quantum theory, he found that it was erroneous to explain properties of solids on the basis of the joint behavior of continuum atoms and molecules as did *the kinetic gas-molecular theory*. Hence, he assumed that solids were composed of individual building elements (atoms and molecules) discretely ordered in crystal lattice structures.[*] Each element has three degrees of freedom and performs 3D oscillations (vibrations) about a fixed position. The entire solid composed of N atoms can be modeled as a multi degree harmonic oscillator with 3N number of motions. All atoms vibrate autonomously with one and the same frequency (v_E) performing longitudinal harmonic oscillations along the three axes of the Cartesian coordinate system. Yet, atom amplitudes depend on atom location within the solid topology (see Figure 1.2).

Note that each individual oscillator (atom or molecule) can change its energy by discrete steps (*quanta*), only. Hence, an *Einstein ideal solid* consisting of N simple oscillators can occupy 3N *quantum states*, only, and all of them are aliquot parts of energy $\hbar\omega_0$ (here $\hbar = h/2\pi$ is the Planck–Dirac constant and ω_0—angular frequency of oscillation). If the index of the energy orbital is zero, this means that the atom does not oscillate. If the index is 1, the atom oscillates with basic frequency ($f_{(1)} = 1 \, h\omega_0$). If the index of the energy orbital is 2, the atom starts oscillating with double basic frequency ($f_{(2)} = 2 \, h\omega_0$), etc. The internal energy of a solid is expressed as a product of the number of excited energy orbitals and the energy of each atom (oscillator)

$$E = 3N_* \overline{n}(\hbar\omega_0) = 3N_*(e^{\beta\hbar\omega_0} - 1)^{-1} \cdot (\hbar\omega_0),$$

where β is the Lagrange multiplier ($\beta = 1/k_\beta T$, k_β—Boltzmann constant).

---

[*] 230 atom configurations are possible in a crystal lattice, and they are arranged in 32 categories being reduced to six basic schemes: isometric, tetragonal, orthorhombic, hexogen, monoclinic, and triclinic.

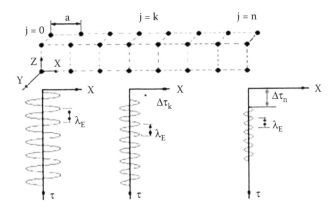

FIGURE 1.2
Longitudinal vibrations of atomic structures (Einstein waves).

The Einstein model has two regimes of operation—under low and high temperature

$$u = \begin{cases} 3Nh\omega_0 \cdot e^{(-\hbar\omega_0/kT)} & \text{– under low temperature} \\ 3N \cdot k_\beta T & \text{– under high temperature} \end{cases} \tag{1.15}$$

Since heat capacity C_v is found from the differential du/dT, it is obvious that $C_v = 3$ Nk_β under high temperature is in accordance with the Dulong–Petit law. Under low temperature, however, C_v is a quadratic function of temperature T and is specified by the expression

$$C_v = 3N\,\hbar\omega_0\,e^{(-\hbar\omega_0/kT)} \approx \frac{1}{T^2}\,e^{(\hbar\omega_0/k_\beta T)}, \tag{1.16}$$

which agrees well with the experimental data. This result of Einstein was taken to be the strongest confirmation of the plausibility of his "ideal solid" model. Hence, Einstein 3D oscillator is treated in modern theory as a legitimate model describing the behavior of a solid, and it is employed in thermophysics and thermodynamics.

The first researcher to improve on the Einstein solid model was Peter Debye.[*] He included additional transversal wave oscillations and showed that the thermal capacity of a solid changes in proportion to T^3 at low temperature (the T^3 model)

$$C_v = 3N_* \left(\frac{\pi^2 \cdot k^4}{5\,C_{max}^3} \right) T^3. \tag{1.17}$$

Debye's model proved to be more useful than the Einstein model and yielded better agreement with the experimental evidence, especially at low temperature.

---

[*] Peter Debye (1884–1966), 1934 Nobel Prize winner in chemistry.

The next improvements of the solid model were due to Enrico Fermi[*] who discovered the property of solids to emit free electrons as an electron gas, the latter playing an important role in energy transfer in solids. It was found that at normal and high temperature, electron (electron gas) contribution to the heat capacity of a working body of a TDS is smaller as compared to the contribution of lattice atoms. Yet, at low temperature, the electron effect on the total heat capacity dominates, that is,

$$C_v = k_e \cdot T + k_l \cdot T^3. \tag{1.18}$$

Here k_e is a coefficient accounting for the electron effect over on heat capacity while k_l expresses the contribution of atom oscillations to the energy accumulated by the lattice of atoms.

The properties of solids to conduct and filtrate energy or hamper energy transfer are of special interest. Their involvement in the increase of energy efficiency of buildings will be discussed in Chapter 5. However, efficiency is an object of a specific technology of design and thermal insulation of structures. It is noteworthy that modern theory treats energy transfer in solids as performed by quasi-particles called phonons, which possess pronounced wave properties and obey Bose–Einstein statistics. Hence, they are treated as belonging to the class of bosons[†] which also include light photons.[‡]

1.1.5 Work Done in a TDS

As clarified, one of the basic results of the interaction between material objects within the control volume of a TDS is the performance of work (mechanical, thermal, chemical, electrical, and electromagnetic). The specific work done in a TDS can be estimated by calculating the scalar product between the generalized force vector $\vec{F}$ and the generalized displacement vector $\vec{dr}$ that comprises the generalized coordinates of the quantities characterizing energy interaction (as is in theoretical mechanics). The scalar product reads

$$\delta L_{M,} = \vec{F} \cdot \vec{dr} = |F| \cdot |dr| \cdot \cos \alpha, \tag{1.19}$$

and holds true for a TDS with two degrees of freedom (thermal and mechanical)—Figure 1.3.

$$(dV > 0 \to \delta L_M > 0).$$

1.1.5.1 Mechanical Work Done on a TDS

The surface force $d\vec{F}$ acting on the boundary of the control area should be collinear and unidirectional with the displacement $\vec{dr}$ ($\angle \alpha = 0$, $\cos \alpha = 1$), respectively. Then, the mechanical work $\delta L'_M$ done for a radial displacement (expansion/compression) of a segment with area dA_0 will be

$$\delta L'_M = \pm dF \cdot dr.$$

---

[*] Enrico Fermi (1901–1954), 1939 Nobel Prize winner in physics.
[†] Bosons (including photons, phonons, pions, gluons, and W and Z bosons) are elementary particles with zero or integral spin.
[‡] You can read more in the book *Energy Modeling and Computations in the Building Envelope*, CRC Press/2015.

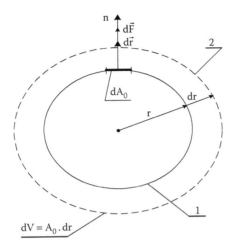

FIGURE 1.3
Mechanical work done on a TDS consists in expansion (compression) of the volume of the control area: 1—initial state and 2—final state.

Yet, since the force dF acting on an elementary surface area dA_0 is $dF = p \cdot dA_0$, the force F acting on the entire control surface of the TDS can be calculated via the surface integral

$$F = \int_{A_0} p \cdot dA_0 = p \cdot A_0,$$

and the mechanical work done for TDS expansion/compression by dr will be

$$\delta L_M = F \cdot dr = p \cdot A_0 \cdot dr = p \cdot dV, \tag{1.20}$$

where $dV = A_0 \cdot dr$ is a control volume with mass M.[*]
    The form of the elementary mechanical work δl_M done on a TDS with unit mass and for $dv = dV/M$ will read

$$\delta l_M = p \cdot dv, \tag{1.21}$$

where

$$\frac{\delta L_M}{M} = \delta l_M \quad \text{and} \quad \frac{dV}{M} = dv.$$

To calculate the work needed to transfer a TDS from initial state 1 (p_1, v_1) to state 2 (p_2, v_2) the respective low of TDS transition needs to be known. This is so, since work depends on the direction of motion. Note that mechanical work depends on the change of TDS pressure p and specific volume v. Then, it is convenient to plot the work in coordinates "p–v."
    As an illustration, Figure 1.4 shows two different paths of TDS transition from state 1 (p_1, v_1) to state 2 (p_2, v_2). The hatched band under plot "b" has the shape of a rectangular

---

[*] Mechanical work is taken to be positive upon expansion of the control volume, and thermal work is taken to be positive when directed from the surroundings to the control volume.

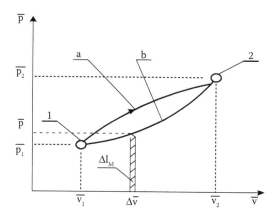

FIGURE 1.4
Mechanical work done on a TDS depends on the process "path": 1—initial state of TDS and 2—final state of TDS. (Notations used are as follows: $\bar{p}$ and $\Delta\bar{v}$—current pressure and increase of the specific volume, while m_p and m_v are scaling moduli specifying the link between the physical quantities p and Δv and their plots $\bar{p}$, mm and $\Delta\bar{v}$, mm.)

trapezoid with area $\Delta l_M = \bar{p} \cdot m_p \cdot m_v \cdot \Delta\bar{v}$ and expresses the work done for TDS "elementary" expansion by $\Delta v = m_v \cdot \Delta\bar{v}$.

Summing all Δl_M we get the following expression for the TDS expansion from state 1 to state 2:

$$l_{M_{1-2}} = \sum \Delta l_M = m_p \cdot m_v \sum_{1}^{2} \overline{p \Delta v}$$

area of the figure $v_1 - 1 - b - 2 - \bar{v}_2$.

Since paths "a" and "b" bound different areas, it is seen that the mechanical work done on a TDS, similar to the mechanical work known from theoretical mechanics will depend on the conditions of $1 \rightarrow 2$ transition (here path "a" bounds an area larger than that bounded by path "b," and hence *work done along the line* 1-a-2 is greater than the work done along the line 1-b-2).

To apply analytical methods in assessing the work done on a TDS, one should use a functional relation of the form $p = f(v) = f(R_\mu T/v)$. Then, the work done to expand the TDS from state 1 to state 2 will read

$$l_{1-2} = \int_{1}^{2} p \cdot dv, \tag{1.22}$$

where the quantity $(R_\mu T/V)$ takes different forms in special cases of thermodynamic transition between two final states of a TDS—for instance (p = const, T = const, v = const).

1.1.5.2 Thermal Work Done on a TDS

Classical thermodynamics assumes that besides the mechanical work done for volume expansion/compression, thermal work in the form of entropy transfer is also done on a TDS with two degrees of freedom.

If a TDS has initial temperature T_1 and entropy S_1 (at state 1) it will occupy equilibrium state 2 with parameters T_2 and S_2 as a result of the changes taking place at its boundaries (see Figure 1.5). It becomes clear that transition from state 1 to state 2 can take place along different paths—for instance 1-a-2 or 1-b-2. Hence, a TDS would transfer different quantity of heat. Besides, the elementary thermal work done will be assessed as[*]

$$\delta L_T = T \cdot dS \quad \text{or per unit mass as} \quad \delta l_T = T \cdot ds, \tag{1.23}$$

and in a finite incremental form it will read

$$\Delta L_T = \Delta Q = T \cdot \Delta S \quad \text{and} \quad \Delta l_T = \Delta q = T \cdot \Delta s.$$

The last formula shows that the change of the thermal work Δl_T is equivalent to the change of heat Δq within the system. The larger the thermal work Δl_T, the larger the accumulated/released heat Δq.

Graphical illustration of the *thermal work* done on a TDS is shown in Figure 1.6. The hatched area is a plot of the elementary thermal work Δl_T done during a process proceeding along line "1-b-2." Similar to the previous paragraph, the thermal work done will depend on the path covered by the TDS between states 1 and 2. The solution of the problem of calculating the quantity of energy accumulated or released by the control volume is found by integrating Equation 1.23 within the corresponding limits

$$q_T = l_{1-2_T} = \int_1^2 T \cdot ds,$$

whereas the analytical function setting forth the temperature change within that interval is specified by the control volume energy balance of (see Section 1.2).

Figure 1.5 shows that the thermal work done on a TDS (i.e., heat accumulated/released by a TDS) depends on the locus of the realized process. For instance, if the process of

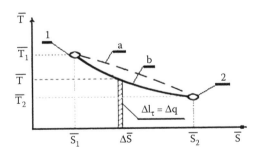

FIGURE 1.5
Thermal work done on a TDS depending on the path of process development: 1—initial state of TDS and 2—final state of TDS.

---

[*] Proposed by Rudolf Clausius (1869r).

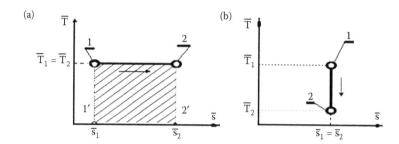

FIGURE 1.6

Thermal work done during two special TDPs: (a) isothermal and (b) adiabatic: 1—initial state of TDS and 2—final state of TDS.

transition from state 1 to state 2 would proceed along line 1-a-2, thermal work would be represented by the area "1′ −1 − a − 2 − 2′." On the other hand, if this process would take place along line 1-b-2, it would be represented by area "1′ − 1 − b − 2 − 2′," respectively.

For the two special cases (isothermal process (T = const) and adiabatic process (s = const)) shown in Figure 1.6a,b, the calculation of the thermal work (heat quantity) is easy and common.

The thermal work done during an isothermal process (Figure 1.6a) is equal to the area of the rectangle 1′ − 1 − 2 − 2′:

$$Q_{1-2_{(T=const)}} = T(S_2 - S_1),$$ (1.24)

and that done during an adiabatic process (Figure 1.6b) is zero:

$$Q_{1-2_{(S=const)}} = 0,$$ (1.25)

since $S_1 = S_2$ and $\Delta S = 0$.

1.1.5.3 Other Types of Work in the System

Except for mechanical work, some modern energy transformers such as fuel cells, batteries, or Seebeck transformers do electrical or chemical work. It is presented in the form of diagrams with characteristic potentials (see for instance Figure 1.7a,b) showing an electric

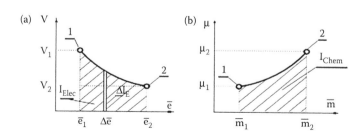

FIGURE 1.7

Plot of the work done by a TDS with multiple degrees of freedom: (a) electrical work and (b) chemical work.

potential and a chemical potential and respective generalized coordinates (electric charge e_0 and mass m). The elementary work done in those cases is given by

- $\Delta Le = V \cdot de0$—electrical work
- $\Delta L_x = \mu \cdot dm$—chemical work

Hatched areas under the plots in Figure 1.7 show the final amounts of electrical and chemical work done within the interval (1–2), respectively.

Note that work done is used to estimate the efficiency of the respective energy-generating device.

Hence, one should account for the area under the plot in Figure 1.7a if the useful effect expected is electric power generation, and for the plot in Figure 1.7b if the useful effect expected is production of a new amount of mass of a chemical compound.

1.1.6 Cascade Mechanism of Energy Conversions and Transformations (Matrix of Direct Transformations)

As known from physics (in agreement with the conservation law[*]), the *quantity of matter* in an insulated system remains constant. This fact can be formally expressed as

$$\sum_n (m_i + E_i) = \text{const.} \tag{1.26}$$

From this fundamental general law (Equation 1.26), follows a law of the conversions of energy into mass and vice versa, laying the foundations of power engineering and heat technics.

Rewriting Equation 1.26 into a sum of finite variations of mass $\sum_n \Delta m_i$ and energy $\sum_n \Delta E_i$ yields an expression for the zero material balance

$$\sum_n \Delta m_i + \sum_n \Delta E_i = 0 \quad \text{or} \quad \sum_n \Delta m_i = -\sum_n \Delta E_i.$$

It stands clear that each mass variation $\sum_n \Delta m_i$ is converted into energy variation $\sum_n \Delta E_i$ and vice versa.

For low velocity of a material object w_i ($w_i \ll w_{Ph} = 3 \cdot 10^8$ m/s) mass variation $\Delta m_i = \Delta E_i / (w_{Ph})^2$ is negligible. For instance, if a body with mass m = 1000 kg and specific heat capacity $C_v = 4186$ J/kg is heated by $\Delta T = 100$ K, decrease of its mass will be insignificant and practically negligible since

$$\Delta m_i = \frac{1000 \cdot 4186 \cdot 100}{(3 \cdot 10^8)^2} \approx (0.4 \cdot 10^{-8} \text{kg}) \Rightarrow 0.$$

Hence, in common engineering practice where mass variation is negligible ($\sum_n \Delta m_i \Rightarrow 0$), the conservation law takes a different form, and it is specified by two separate equalities:

$$\sum_n m_i = \text{const} \quad \text{and} \quad \sum_n E_i = \text{const.} \tag{1.27}$$

---

[*] Postulated by Lermontov (1748).

Both Equations 1.27 show that inside an insulated material system, such as a TDS, *mass and energy can be neither created nor destroyed but remain constant in time.* The second Equation 1.27 may comprise different energy components, that is,

$$\sum_n E_i = E_M + E_{Th} + E_{El} + E_{El.M} + E_{Chem} + E_{Nucl} = \text{const.} \qquad (1.28)$$

Equation 1.28 says that although the total amount of energy in a TDS is constant, one type of energy can be converted into another. For instance, chemical energy (E_{Chem}) can be converted into heat (E_{Th}), mechanical energy (E_M)—into thermal energy, and vice versa, chemical energy (E_{Chem})—into electrical energy (E_{El}). These types of physical phenomena are observed by the century-old practice of civilization. Such form of the law of energy conversions is known as the *First Law of Thermodynamics.*[*]

As noted in the introduction, a number of physical processes of energy conversion exist. They are ordered in the so-called "Matrix of direct conversions" (see Table A.1 in the Appendices). It specifies 25 processes and devices that convert energy from one type into another. For example, the first row of the matrix specifies all devices (media) where mechanical energy is converted into thermal, electrical, chemical, and radiation (luminous) energy. The so-called energy converters are located along the diagonal where only energy "parameters" change but energy form is conserved.

The modern use of energy implies the following sequence of energy transformation:

- Extraction of energy carriers and their transportation (first stage)
- Conversion of the energy carriers into a commercial energy product (second stage)
- Use of energy products in energy converters to satisfy modern energy needs (third stage)

As assumed, each stage of energy "conversion" into a consumable commodity is specifically named as shown in Figure 1.8, and the coverage of the successive stages is called "energy cascade."

The final energy consumption is measured during the third stage when assessing the energy spent for the operation of modern equipment, transport means, and industrial installations. To prepare the energy balance of companies and branches of national economy, the ultimate consumer (UC) should refer to the table accounting for the four trends—industry, transport, service, and household. However, consumption of primary energy can be followed only after designing a scheme of a UC energy cascade.

An example of cascade transformations of primary energy and an energy model of central electricity supply are shown in Figure 1.9. Lignite coal is a carrier of primary energy. It is mined, transported to a thermal electric power station, dried and milled into final coal powder, losing 5%–10% of its primary energy. Next, coal powder burns in power boilers releasing radiation and hot combustion products. If the power station is of steam-condensation type, the Rankine cycle of energy generation is realized (Section 2.2). In its turn, the steam produced is used to set in motion steam turbines aggregated with AC generators. De-energized steam condenses after leaving the turbine, and it is collected in condense reservoirs to subsequently evaporate in the cauldrons.

Energy efficiency of the above thermoelectric technology does not exceed 32%–34%. Electricity generated is transferred through a transmission network to the UCs. Hence,

---

[*] Known also as the "First principle of thermodynamics."

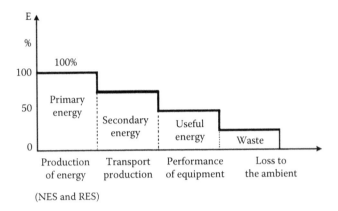

FIGURE 1.8
Scheme of a cascade of energy conversion.

the amount of energy of 10%–25% is lost since electric substations, transmission lines, and transformers/distributors cause additional losses. UCs such as industry, transport, electric cooking appliances, refrigerators, illuminator fixtures, TV and radio sets, CD players, communications (including the Internet), etc. get barely 15%–20% of the total primary energy.

The difference between the used primary energy capacity of the fuel at ES and the energy reached the end users amounts to 80%–85%, and together with wastes of combustion

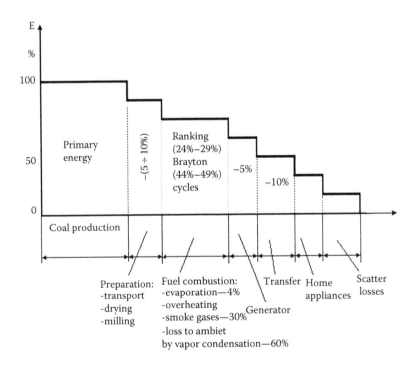

FIGURE 1.9
Energy model of central electricity supply: Rankine cycle operating with efficiency of 24%–29% and Brayton cycle gas-turbine stations operating with efficiency of 44%–49%.

(carbon and nitrogen oxides) pollute the atmosphere, increase the greenhouse effect, and change the global climate. Disastrous effects on ecosystems, water, soil, and human health are observed at the regional level, where such energy generators operate. Hence, huge power stations are the major polluters of air, soil, and water. They should be put under strict and permanent control* and additionally taxed.

Primary energy is from the Sun and processes that take place on its surface. An energy source of secondary importance is the Earth itself with the processes taking place in its core. Solar and geothermal energy, together with the energy of the Moon and Space, are referred to as renewable energies and constitute the group of renewable energy sources (RES). Wind energy, that of tides, bioenergy, gravity (osmotic) energy, and solar radiation are all effects of the RES "performance" and are also popular as renewable energies. Yet, power technologies based mainly on oxidation (combustion) of bioresources (biomass, oil, coal, and natural gas) and nuclear decay are primary energy sources in modern civilization. Those technologies dominate in the energy balance of the world economy.

Fossil fuels are significant energy carriers, but their reserves are limited worldwide, and are facing exhaustion in the near future. For instance, oil is expected to last about 20–30 years, shale gas—10–15 years, reserves of uranium—30 years, natural gas—50–70 years, frozen methane—100–150 years, and coal—150–200 years. Those carriers belong to the group of nonrenewable energy sources (NES). According to recent documents of the EU and the EC, the European Community is broadening its energy generation options by focusing on small energy-generating capacities and use of renewable energy sources. Moreover, the development of intelligent networks of small buildings with zero energy balance and other energy strategies is underway in contrast to large-scale and inefficient energy-generating capacities.

The technical progress of modern civilization is measured not only by the degree of life "energizing," human activity and the quality of power technologies, but also by the "depth" of the industrial stamp left on the planet. Without strict public control in people's and the environment's interest, the technical progress and energy voracity of modern civilization will result in local dependence on the import of energy and resources, possibly resulting in global economic and ecological instability.

1.1.7 Control Questions

1. What is the result of the interaction between material bodies and energy fields?

2. Which are the generalized forces? Which are the generalized coordinates/displacements?

3. Which are the components of a TDS?

4. How could one find the degrees of freedom of a TDS?

5. How could one calculate the mechanical work done on a TDS? How could one visualize it graphically? Which is the SI unit for mechanical work?

6. How could one calculate the thermal work done in a TDS? Which is the coordinate system to plot the thermal work in? Does the heat quantity depend on the path of the energy-exchange process in a TDS?

7. Specify at least three processes converting electrical energy into another type of energy.

---

* Since 2008, there are new regulations of calculating greenhouse gas emissions. A new system of greenhouse gas trading will be introduced in 2030.

8. Which are the characteristics of the ideal gas?

9. Which are the specific features of the Einstein solid?

10. Which scientific fact unleashed a crisis in the classical theory of heat? How was this crisis overcome?

11. What is the link between the increase of internal energy and increase in temperature?

12. What is the interrelation between change of enthalpy (total energy) and temperature?

13. What is an energy cascade?

14. How can one calculate the ultimate energy consumption?

15. What does the law of energy conversion state?

16. What is the link between mass change and energy change in a closed TDS?

17. Which are the energy sources known as RES?

18. Specify the estimations of the availability of fossil energy sources.

1.2 Basic Principles of Thermodynamics

Thermodynamics is an engineering branch given birth to in the 18th century by the works of M.V. Lomonosov (1711–1765) on the kinetic theory of matter, where the first object of study covered all forms of energy conversions. Yet, classical thermodynamics really began with the publication of the fundamental treatise of Nicolas Léonard Sadi Carnot (1796–1832) on fire and engines and by the works of Rudolf Clausius (1822–1888) and James Maxwell (1831–1879). Modern thermodynamics emerged at the end of the 19th century, being stimulated by the results of Ludwig Boltzmann (1844–1904), Max Plank (1858–1947), Albert Einstein (1879–1955), and many other researchers.

As already discussed, the fundamental laws of physics of mass conservation and energy conversions are valid in thermodynamics and heat transfer, too (see Section 1.1.6). Hence, we shall consider in what follows their validity for various TDS (including noninsulated and open ones) where different processes (isochoric, isobaric, isothermal, or adiabatic) occur.

1.2.1 Law of Energy Conversions (First Principle of Thermodynamics)

As shown above, the law of energy conversions under normal conditions ($c_i \ll 3 \cdot 10^8$ m/s) is expressed by Equation 1.28. If a TDS is insulated and closed, it is assumed to have "internal" kinetic energy is conserved, that is,

$$E = U = \text{const.}$$

Rewriting the above expression for a TDS with a unit mass, it takes the form: $U/m = u = \text{const}$ or $C_v dt = \text{const}$. Yet, real TDS are not insulated and open.

If the real TDS exchange energy with the environment in the form of heat Q and/or mechanical work L_M. Then, the law of energy conversion will read

$$U + L_M - Q = 0, \tag{1.29}$$

where TDS internal energy is equal to U, and the sum $\Sigma E_i = E_{Th} + E_M = -Q + L_M$ is interpreted as that of external energy impacts on a system with two degrees of freedom (note that Q > 0 when the heat flux is directed to the TDS, and the mechanical work is positive when the TDS expands).

For a TDS with unit mass, Equation 1.29 takes a new form

$$q = u + l_M, \tag{1.30}$$

and shows how the thermal energy is converted into mechanical and internal energy, *and vice versa.*

Here $u = U/m$, $l_M = L_M/m$, and $q = Q/m$ are TDS internal energy, mechanical work, and heat. Adopting finite increments (Δq, Δu, Δl_M) and differentials (δq, du, δl_M) a new form of the law is found

$$\Delta q = \Delta u + \Delta l_M \tag{1.31}$$

and

$$\delta q = du + \delta l_M. \tag{1.32}$$

This means that heat Δq (δq) introduced in the TDS is consumed to change the TDS internal energy Δu (du) and/or TDS volume.

The rewritten form of Equation 1.32 $\delta l_M = \delta q - du$ shows that change of the TDS volume (dv $\Rightarrow$ var) yields change of the quantity of internal energy (du) and/or change of the quantity of introduced (released) heat (δq).

The substitution of Equation 1.21 in 1.32 yields the following equation for the TDS thermal balance:

$$\delta q = du + pdv = dh - vdp \tag{1.33}$$

note that enthalpy h is set forth by the equality $h = u + pv^*$ and $dh = du + d(p \cdot v) = du + p \cdot dv + v \cdot dp$. The equation of the energy balance for TDS with mass m and for the whole control area takes the form

$$\delta Q = dU + pdV = dH - Vdp \quad (H = hm). \tag{1.34}$$

When a TDS is in an initial state "1" characterized by ($p_1 v_1 h_1 u_1 T_1$), it occupies state "2" due to the introduced heat $\delta q > 0$. The TDS new state parameters ($p_2 v_2 h_2 u_2 T_2$) can be predicted integrating Equation 1.33, for example:

- The value of the internal energy u_2:

$$u_2 = u_1 + q_{1-2} - \int_1^2 pdv$$

---

* Introduced by J.W. Gibbs (1839–1903) in 1875r. For a unit mass, the enthalpy (ἔνθαλπος) is $h = u + p \cdot v$.

- The value of enthalpy h_2:

$$h_2 = h_1 + q_{1-2} + \int_1^2 v \, dp.$$

It is seen from the above relations that energy (u_2) and enthalpy (h_2) of the new state 2 are determined by the introduced quantity of heat (q_{1-2}) and by the values of integrals $\int_1^2 p \cdot dv$ and $\int_1^2 v \cdot pd$, depending on the "paths" of the mechanical work done (this problem will be discussed in what follows).

It follows from those arguments that the application of the *law of energy conversions* in a noninsulated TDS enables one to predict the TDS behavior in a final equilibrium state (see Chapter 2) under certain finite external impacts (introduced/released heat (q_{1-2}) or done mechanical work ($l_{M_{1-2}} = \int_1^2 p \cdot dv$)).

When a TDS is also open, besides heat (δq) and mechanical work ($\delta l_M = p \cdot dv$), one should also add to the external impacts the energy (wdw) of mass introduced into the control volume. The principal scheme of that system is shown in Figure 1.10, and two laws of conversion simultaneously operate in it (see Equation 1.27).

1. Law of mass conservation (continuity law):

$$m^* = \rho \cdot A_i \cdot w_i = \frac{w_i \cdot A_i}{v} = \text{const}, \tag{1.35}$$

where

- $m^* kg/s$—mass flux
- $\rho = v^{-1} kg/m^3$—density
- $A_i m^2$—area of the i^{-th} "open" cross section of the control volume
- $w_i m/s$—velocity within the ith cross section of the control volume

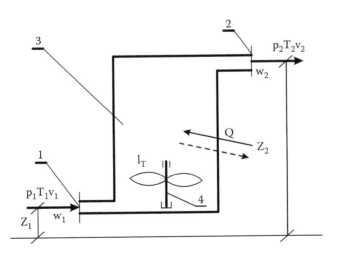

FIGURE 1.10
Scheme of an open TDS: 1—intake (p_1, v_1, w_1, T_1); 2—exhaust (p_2, v_2, w_2, T_2); 3—working volume (coincides with the control volume); and 4—fan (pump).

Note that written in the above form, the continuity law (i.e., that of mass conservation) requires the maintenance of constant mass flux (m* = const) through each *i*th cross section (including sections 1 and 2 of the TDS entrance and exit).

The following form of Equation 1.35 is valid for the entrance and exit of a TDS having an incompressible working fluid (v = const):

$$\frac{w_1 \cdot A_1}{v} = \frac{w_2 \cdot A_2}{v}$$

and its solutions in terms of velocity w_2 or area A_2 at the TDS exit read $A_2 = A_1(w_1/w_2)$ and $w_2 = w_1(A_1/A_2)$, respectively.

The relations thus found can be used to assign dimensions to the physical device or assess the hydraulic losses upon leakage of the working fluid out of the energy converting appliance, knowing w_1 and A_1.

2. *Law of energy conversions in open TDS*: The second conversion law, in contrast to a noninsulated system, provides information about the distribution of additional energy fluxes which are internal for the TDS. It reads for open systems[*]

$$q_{1-2} = (u_2 - u_1) + \int_1^2 pdv + \left(\frac{w_2^2 - w_1^2}{2} \right) + g(z_2 - z_1). \qquad (1.36)$$

Introducing m*kg/s, it takes the form

$$Q_{1-2} = (U_2 - U_1) + \int_1^2 pdV + \frac{m^*}{2}(w_2^2 - w_1^2) + m^* g(z_2 - z_1)$$

and in differential form, it reads

$$\delta q = du + pd(v) + wdw + gdz \qquad (1.37)$$

or

$$\delta q = dh - vd(p) + wdw + gdz. \qquad (1.38)$$

The new terms in Equations 1.36 through 1.38 are those which account for (i) the change of the kinetic energy due to working fluid motion in the TDS [$0.5(w_2^2 - w_1^2)$ and wdw] and (ii) the change of the potential energy due to the displacement and gravitation and [$g(z_2 - z_1)$ and gdz] between entrance 1 and exit 2. They are included in the equation of energy conversions in order to consider the specific structural characteristics of the open TDS. The quantitative analysis of the individual tribute of those terms to the equation under consideration may result in improvement of the quality of the energy device. For

---

[*] Work done to overcome internal friction forces (l_{mp}) and work done to fill/discharge the system ("technical work"—l_T) are disregarded.

example, the efficiency of using the imported heat (q_{1-2}) can be estimated via the value of the internal energy u_2 at the exit cross section 2

$$u_2 = u_1 + q_{1-2} - \int_1^2 pdv - 0.5(w_2^2 - w_1^2) - g(z_2 - z_1).$$

The new form of the energy equation shows to what extent its negative terms expressing internal energy conversion reduce the effect of imported heat (q_{1-2}).

We may conclude that despite the advantages outlined, the energy balance equations of an open TDS are as informative about the running processes as those valid for an *insulated TDS*. To perform analysis of the power technologies, however, *we shall consider* in the present book simpler forms of the balance equations (we shall mostly use Equations 1.33 and 1.35 in what follows).

1.2.2 Law Concerning the Natural Direction That a TDP Follows (Second principle of thermodynamics)

The necessity of such a law emerged out of the design of engines where heat was converted into mechanical work. Starting from the 17th and 19th centuries, many inventors and designers proceeded to design (even nowadays) the "perpetual motion machine" (i.e., an infinitely operating engine) using a single energy source—Earth (gravity) or Sun (solar energy).

Carnot, Clausius, Thompson, Plank, Boltzmann, and other researchers proved the inconsistency of such projects formulating the second principle of thermodynamics. In a classical formulation, this law states that the cyclic functioning of an engine needs the simultaneous availability of *two heat sources* (bodies)—warm and cold. *The operation of an engine that gets thermal energy from one source (body) only is impossible.*

This is illustrated by Figure 1.11. Assume at first that a TDS contacts a hot body only, receiving heat Q. Its internal pressure increases and the working fluid expands displacing the moving wall. This process proceeds until the moving wall lifts to height Z_2 (Figure 1.11b). Then, fluid expansion stops since pressure drops to a degree where equilibrium between active forces reduced to the moving wall is attained (i.e., $\Sigma F_{iz} = 0$), and the equilibrium is specified by the condition $pA_0 = G$ or $p = G/A_0$. If the TDS is insulated from a cold source, load G will *remain in the final top position* $(Z = Z_2)$—Figure 1.11b). A repeating mechanical cycle occurs when the moving wall subjected to force $\overline{G}$ drops from position $(Z = Z_2)$ to its down position $(Z = Z_1)$, then again lifts to position $(Z = Z_2)$, etc. The wall would drop to position $(Z = Z_1)$ if the internal pressure (p) would drop to its initial value thus decreasing the elevating power $(p_2 . A_0)$.

Technically, this can be realized in two ways via:

- Cooling of the closed TDS as illustrated in Figure 1.11c
- Opening the TDS and outflow of the working fluid into the cold environment as shown in Figure 1.11d

Hence, the TDS contacts a hot body (a high-energy carrier), expands, and performs positive work $(l_{M_{1-2}})$—the movable wall shifts from position Z_1 to position Z_2. To perform a cycle, however, the TDS should interact with a cold body (low-energy source) releasing heat in order to constrict and thus doing negative work (the load drops from position Z_2 to position Z_1). Hence, the thermal energy in both phases of the cycle is transferred from

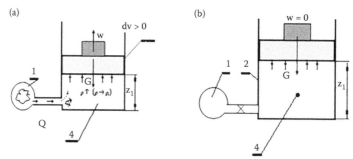

TDS with a single energy source (hot body)—the moving wall remains in top dead center.

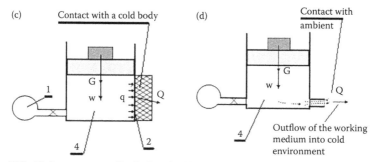

TDS with 2 energy sources (hot body and cold body)—the moving wall returns
at its initial position (lower dead center).

FIGURE 1.11
Illustration of the second law of thermodynamics: (a) TDS expands by $dQ > 0$, the load elevates; (b) TDS is in equilibrium ($G - p \cdot A_0 = 0$), the load is in top dead center; (c) *first variant*—TDS constricts (heat is released to the cold body 3); and (d) *second variant*—TDS compresses by exhausting the combustion products 4 into the cold environment and pressure drops ($p \cdot A_0 = G$): 1—hot body; 2—TDS control surface; 3—cold body (environment); and 4—combustion products; $\vec{G}$—force reduced to the moving wall.

areas with high temperature to areas with low temperature (which is the natural course of transfer).

Theoretically, the second principle of thermodynamics proves that if a TDS is in contact with a single energy source, it expands attaining equilibrium. This will occur when pressure (p) generates a force ($p_0 \cdot A_0$) large enough to equalize with the opposing force G. Then the top dead center of the movable wall will be reached and the mechanical system will stop permanently. Its motion will not be repeated periodically.

Hence, the natural course of a process occurring in a TDS is to increase entropy s by increasing the volume of the control area, the number of thermodynamic states, and the thermodynamic probability.[*]

There are over 20 formulations of the second principle of thermodynamics.[†] The shortest one states that entropy s of any insulated system increases to attain maximal value s_{max} under specific equilibrium conditions.[‡] The opposite process is impossible.

---

[*] $S = k_B \ln W_T$, proposed by L. Boltzmann in 1906. W comes from "Wahrscheinlichkeit"—the frequency with which a TDS would occupy a certain macrostate.

[†] Formulations of Carnot (1796–1832), Clausius (1822–1888), Thompson (1824–1907), Caratheodori (1873–1950), and others.

[‡] At present, "the minimum of the Potential of free energy" is used instead of "the maximum of the Entropy" (see Dimitrov A.V. (2015)).

Entropy of an equilibrium state $s_{max} = s_2$ can be calculated by

- Using the classical formula of Clausius $s_2 = s_1 + q_{1-2}/T$, where q_{1-2} is heat received by the hot body (reduced to unit mass) under temperature T or
- Finding the thermodynamic probability of Boltzmann (see in Dimitrov A. V. 2015)

In conclusion, the two basic laws of thermodynamics provide engineering knowledge on the mechanisms of energy conversion (the first law) and on the running direction (the second law) of energy processes in thermal machines—engines and converters. The organization of the interaction between TDS and hot and cold energy sources will be discussed in Section 1.3.

1.2.3 Control Questions

1. What are the consequences of introducing environmental heat into a TDS?
2. Is the thermal energy positive or is it negative when a TDS releases heat? What are the consequences of this effect?
3. What does happen when a TDS constricts? What is the sign (+ or −) of the mechanical work?
4. What does the law of energy conversion applied to a noninsulated TDS state?
5. How can one predict the new TDS parameters after heating or cooling a TDS in equilibrium? Which is the quantity that specifies parameter new values?
6. How can one analytically express the energy conversions law applied to an open and noninsulated TDS?
7. What is the formulation of the continuity law?
8. What is the difference between the balance equations of the first principle of thermodynamics of a noninsulated TDS and the respective balance equations of an open TDS? Is there qualitative difference between them?
9. Why is the design of a periodically operating heat engine with a single energy source impossible?
10. Can a heat engine function if it is simultaneously linked to the two sources?
11. Is the sequence of contacts between a TDS and the energy sources important for the engine operation?
12. What does the second principle of thermodynamics state?

1.3 TDPs in an Ideal Medium

Modern science assumes that all changes of the state of a TDS resulting from various energy interactions are TDPs. Since those changes are invisible and intangible, they can only be established via physical measurement of the basic state parameters: temperature T, pressure p, volume v, electric potential, etc. If the parameters $(T_1 p_1 v_1)$ of a TDS at state "1" attain stable values $(T_2 p_2 v_2)$ at state "2", that is, state parameters attain constant (equilibrium) values at the new state, the transition process "1-2" is called equilibrium process. If as a result of energy interactions no equilibrium at the final state "2" can be attained, the process is called nonequilibrium process.

When a certain equilibrium process "1-2" runs in an ideal medium (disregarding the effects of intermolecular forces and vortex effects operating in an open TDS), it can be reversible (this means that no external impacts on the TDS are needed to return the system to its initial state). Yet, no spontaneous reversible processes occur in nature, since they contradict the natural course of energy conversion (where entropy tends to maximum). In what follows we shall consider equilibrium TDPs, only, which are described by the Clausius–Clapeyron equation (see Equation 1.4).

To assist readers, we shall treat at the beginning the so-called *elementary TDPs* imposing parametric restrictions on them (one of the state parameters v, p, T, s is fixed and does not vary). The analysis of those processes elucidates the direction of TDS restrictions. Yet, elementary processes will be used as "building" blocks of the thermodynamic cycles—see Section 1.4 and Chapter 2 that follow. To start the explanation, we shall consider a process in a TDS with fixed boundaries and constant volume (the so-called "isochoric" process).

1.3.1 Isochoric Process (Charles Process)

Consider a noninsulated closed TDS with constant volume $(v = \Rightarrow dv = 0)$—Figure 1.12. Then, the Clausius–Clapeyron equation reads

$$\frac{p}{T} = \frac{R_\mu}{v} = const \quad or \quad p = const * T. \tag{1.39}$$

Hence, it follows that pressure variation dp and temperature variation dT are directly proportional to each other, that is, $dp \approx dT$, and they are collinear (i.e., if $dp > 0$, than $dT > 0$, too, and vice versa). To assess the direction of a process running in a TDS, which receives external heat (i.e., $(\delta q > 0)$), we should take into account the following considerations.

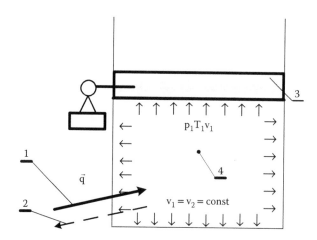

FIGURE 1.12
TDS volume does not vary, mechanical work is zero. 1—TDS heating; 2—TDS cooling; 3—unmovable wall; and 4—control volume.

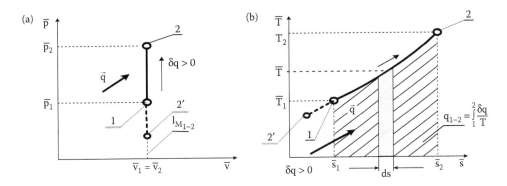

FIGURE 1.13
TDS isochoric heating and cooling does not result in mechanical work ($l_{M1-2} = 0$): (a) mechanical diagram and (b) thermal diagram: 1—initial state; 2—final state at heating of TDS; and 2'—final state at cooling of TDS.

Use the law of energy conversion. Assume $dv = 0$. Then, Equation 1.33 takes the form

$$\delta q = du + p \cdot 0 = du \quad \text{or} \quad \delta q = C_V dT. \tag{1.40}$$

The new form shows that the TDS does not perform work in that case, that is, the mechanical work done is zero ($l_{M1-2} = 0$) since the movable wall remains in its initial position. The external heat δq changes the TDS internal energy, only, and increases the TDS temperature (i.e., for $\delta q > 0 \Rightarrow du > 0$ and $dT > 0$, T since $du = C_V T$).

Summarizing the above arguments, we may conclude that if heat is transferred to a TDS under isochoric conditions, TDS temperature rises proportionally to the heat transferred. Since dT and dp vary collinearly, pressure will also increase ($dp > 0$) as illustrated in Figure 1.13a—see the plot of the mechanical work ($\bar{p} - \bar{V}$). If the initial state "1" of a TDS is specified by the two parameters $\bar{p}_1$ and $\bar{V}_1$, the final state "2" lies $\bar{p}_2$ is found from Equation 1.39

$$p_2 = p_1 \frac{T_2}{T_1}.$$

The diagram of the thermal work ($\bar{T} - \bar{s}$)—Figure 1.13b shows that the TDS initial state (point "1") is specified by T_1 and s_1 (by their plots $\bar{T}_1$ and $\bar{s}_1$, respectively).

Since heating of the TDS yields $\delta q = du$ and $q_{1-2} = u_{1-2} = C_v (T_2 - T_1)$, the final point "2" will have coordinates T_2 and s_2, supplied by the expressions for temperature and entropy

$$T_2 = T_1 + \frac{q_{1-2}}{C_v}, \tag{1.41}$$

$$s_2 = s_1 + \int_1^2 \frac{\delta q}{T} = s_1 + \int_1^2 \frac{du}{T} = s_1 + C_v \int_1^2 \frac{dT}{T} = s_1 + C_v \ln \frac{T_2}{T_1} \tag{1.42}$$

or

$$s_2 = s_1 + C_v \ln \frac{p_2}{p_1}.$$

If a TDS releases heat ($\delta q < 0$) processes are opposite, namely pressure in the diagram $(\overline{P} - \overline{V})$ drops along line $\overline{V}_1 = \overline{V}_2'$ to point 2' (see Figure 1.13a). Temperature $\overline{T}_2$ and entropy $\overline{s}_2$ decrease in the diagram $(\overline{T} - \overline{s})$ since the internal energy u_{1-2} decreases and TDS loses heat (see Figure 1.13b).

1.3.2 Isobaric Process*

The isobaric process requires fixed TDS pressure

$$p = \frac{G}{A_0} = \text{const.}$$

The physical conditions of that process are attained in a chamber with constant cross section (position 1) equipped with a moving wall (position 2) and for $G = \text{const}$—Figure 1.14.

Clausius–Clapeyron equation reads

$$\frac{v}{T} = \frac{R_\mu}{p} = \text{const} \quad \text{or} \quad v = \text{const} * T, \tag{1.43}$$

and with respect to the initial and the final equilibrium state of the process, it takes the form

$$v_2 = v_1 \frac{T_2}{T_1}. \tag{1.44}$$

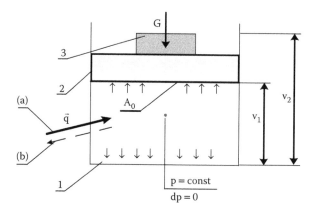

FIGURE 1.14
Wall movability yields $p = \text{const}$: (a) heat supplied to the TDS ($\delta q > 0$) and (b) heat release from the TDS ($\delta q < 0$): 1—control volume; 2—movable wall; and 3—mechanical load.

---

* Formulated in 1802 by Joseph Gay-Lussac and named after him.

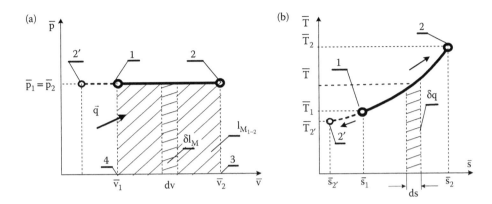

FIGURE 1.15
TDS isobaric heating increases TDS enthalpy and temperature. The system expands along line "1 → 2" performing positive work: (a) mechanical diagram and (b) thermal diagram: 1—initial state; 2—final state at expansion of TDS; and 2'—final state at compression of TDS.

The equation of TDS state (Equation 1.43) shows here that the specific volume v and temperature T are directly proportional to each other (v ≈ T) and collinear. If for instance dT > 0, then dv > 0 and the system expands, that is, v and hence, positive mechanical work $l_{M_{1-2}}$ is done.

The process that occurs when heat is supplied ($\delta q > 0$) isobarically is given by the plot $\bar{p} - \bar{v}$ (Figure 1.15a). The initial position "1" of the system is specified by the couple $(\bar{p}_1; \bar{v}_1)$. To assess the direction of process development after TDS "heating," we use the equation of energy conversion in the form Equation 1.33

$$\delta q = dh - v * 0 = dh \Rightarrow (dp = 0). \tag{1.45}$$

It proves that the whole amount of heat supplied is spent to increase the *enthalpy* (the TDS total internal energy), only. Temperature rises proportionally to the enthalpy (dh = $C_p \cdot$ dT) and TDS is heated. Due to the direct proportionality between v and T, TDS expands at T↑ and $v_2 > v_1$. The process runs along the horizontal line "1-2" (Figure 1.15a).

As shown in Figure 1.15a, the mechanical work done during TDS expansion from point "1" to point "2" is graphically illustrated by the rectangle "1-2-3-4" whose area is

$$l_{M_{1-2}} = \int_1^2 p dv = p_1(v_2 - v_1) = p_2(v_2 - v_1). \tag{1.46}$$

The work is positive for $v_2 > v_1$, which proves that the TDS performs work on the "opposing" the environment and an equal amount of the heat q_{1-2} supplied.

Line "1-2" plots adiabatic heating in the thermal diagram (1.15b), where the coordinates of the equilibrium state "2" are found via the expressions

$$T_2 = T_1 + \frac{q_{1-2}}{C_p}, \tag{1.47}$$

$$s_2 = s_1 + \int_1^2 \frac{\delta q}{T} = s_1 + C_p \int_1^2 \frac{dT}{T} = s_1 + C_p \ln \frac{T_2}{T_1} \qquad (1.48)$$

or

$$s_2 = s_1 + C_p \ln \frac{v_2}{v_1}.$$

Since the heat supplied q_{1-2} is found, the calculation of temperature T_2 ($T_2 > T_1$) is easily performed.

When the inverse problem is posed and the quantity of "supplied" heat q_{1-2} needed to attain temperature T_2, is to be found, one should use expression

$$q_{1-2} = C_p(T_2 - T_1), \qquad (1.49)$$

whereas

$$\Rightarrow T_2 = T_1 + q_{1-2}/C_p.$$

Figure 1.15a illustrates also heat release (TDS isobaric cooling). The process runs from point "1" to point "2‴" and is specified by decrease of TDS enthalpy h (when ($\delta q < 0$), then (dh < 0) and ($h_2 < h_1$)—Equation 1.33). Decrease of h is due to decrease in temperature T (if dh < 0, then dT < 0) and TDS compression takes place (dv < 0)—Figure 1.15a. Temperature drops from T_1 to $T_{2'}$, and the specific volume—from v_1 to $v_{2'}$.

TDS cooling ($\delta q < 0$) results not only in negative mechanical work ($l_{M_{1-2}} = p_1(v_{2'} - v_1) = -p_1(v_1 - v_{2'})$, since $v_{2'} < v_1$), but also decreases entropy ($s_{2'} < s_1$), as seen from the expression $s_{2'} = s_1 + \int_1^2 (-\delta q)/T = s_1 - C_p \ln(T_1/T_{2'})$, or $s_{2'} = s_1 - C_p \ln(v_1/v_{2'})$, and illustrated in Figure 1.15b.

1.3.3 Isothermal Process*

This process runs under constant temperature (T = const, respectively, dT = 0—see Figure 1.16) is considered next of special importance in modeling the cyclic processes of heat engines. Here, the equilibrium state parameters are set forth by the modified equation of Clausius–Clapeyron

$$p \cdot v = R_\mu \cdot T = \text{const} \quad \text{or} \quad \frac{p_1}{p_2} = \frac{v_2}{v_1} \qquad (1.50)$$

proving that TDS volume (v) is inversely proportional to pressure (i.e., if dp > 0, then dv < 0 or if $p\uparrow \Rightarrow v\downarrow$ and vice versa).

---

* Described by the law derived by R. Boyle (1662r.) and E. Mariotte (1676r.).

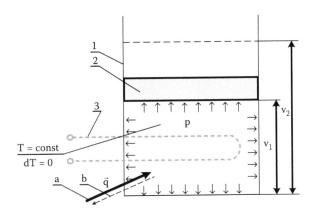

FIGURE 1.16
TDS isothermal expansion and compression take place at the expense of supplied/released heat: 1—control volume; 2—movable wall; and 3—heating/cooling coil.

On the other hand, the analysis of the balance equation of TDS energy per unit mass (Equation 1.33) shows that supplied heat q_{1-2} is used to do mechanical work, only, which follows from the equality:

$$\delta q = C_v \cdot 0 + \delta l_M = \delta l_M \qquad (\Rightarrow dT = 0) \quad \text{or} \quad \delta q = \delta l_M. \tag{1.51}$$

If $\delta q > 0$, work $l_{M_{1-2}}$ is positive and TDS expands—$dv > 0$ ($v_2 > v_1$), pressure drops (i.e., $p{\downarrow}$) due to the inverse proportionality (Equation 1.50). The TDP is illustrated in Figure 1.17a. Starting from point "1" the system expands to point "2" when heat is supplied, and pressure drops from p_1 to p_2.

The mechanical work done can be calculated via

$$l_{M_{1-2}} = \int_1^2 p \cdot dv = \int_1^1 \frac{R_\mu T}{v} \cdot dv = R_\mu T \ln \frac{v_2}{v_1} = R_\mu T * \ln \frac{p_1}{p_2} \tag{1.52}$$

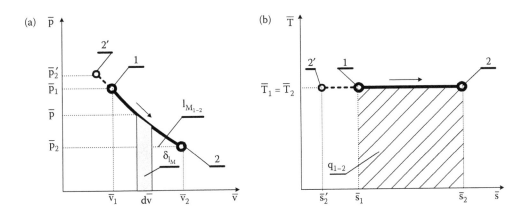

FIGURE 1.17
Mechanical work is done without change of the TDS internal energy (du = C_v * dT = 0), and heat generated is spent for TDS expansion, only. (a) Mechanical diagram and (b) thermal diagram: 1—initial state of TDS; 2—final state of the TDS at expansion; and 2'—final state of the TDS at compression.

and since the integration of Equation 1.51 yields $l_{M_{1-2}} = q_{1-2}$, we find

$$l_{M_{1-2}} = T_1 * (s_2 - s_1) = T_2 * (s_2 - s_1). \tag{1.53}$$

As noted, the external heat supplied q_{1-2} is entirely converted into mechanical work $l_{M_{1-2}}$, since the internal energy remains constant ($du = 0$). As shown by the thermal diagram ($\bar{T} - \bar{s}$—Figure 1.17b), the isothermal process runs along the horizontal line ($T_1 = T_2 = const$), starting from point "1" ($\bar{T}_1, \bar{s}_1$) and directed to point "2" ($\bar{T}_2, \bar{s}_2$), whose coordinates are found as

$$T_2 = T_1, \tag{1.54}$$

$$s_2 = s_1 + \frac{q_{1-2}}{T_1}. \tag{1.55}$$

To solve the inverse problem, when the energy supplied to the TDS and needed to attain a certain level of entropy s_2 is to be calculated, one should use the modified Equation 1.55. When a TDS undergoes isothermal cooling ($\delta q < 0$), the mechanical work done is negative ($\delta l_M < 0$) and TDS constricts ($dv < 0 \Rightarrow v_{2'} < v_1$). The new state corresponds to point "2'" on the plots in Figure 1.17. Besides entropy decrease ($s_{2'} < s_1$), the isothermal cooling is characterized by pressure rise in the TDS ($p_{2'} > p_1$), too. Hence, the isothermal process and the basic analytical relations predicting the change of TDS parameters are useful in (i) modeling heat exchange in various heat-exchangers with and without phase transformations and (ii) approximating the running thermodynamic cycles as discussed in Sections 1.4 and 1.5.

1.3.4 Adiabatic Process

Restrictions are that the TDS should be thermally insulated and closed as illustrated in Figure 1.18. Hence, the equation of energy balance (Equation 1.33) will read here

$$0 = du + \delta l_M = 0 \Rightarrow (\delta q = 0) \quad \text{or} \quad \delta l_M = -du, \tag{1.56}$$

proving that mechanical work (δl_M) is done at the expense of TDS internal energy (u). As a result of expansion, for instance, TDS temperature drops. If the environment performs work on the TDS ($\delta l_M < 0$) and the system constricts, then the internal energy (u) and temperature increase.

Equation 1.51 can be written in the form

$$\frac{du}{\delta l_M} = -1,$$

and modified via routine transformations into two forms[*][†]:

$$pv^k = const \quad \text{or} \quad p_1 v_1^k = p_2 v_2^k, \tag{1.57}$$

---

[*] The substitution of du and δl_M yields $C_v \cdot dT/p \cdot dv = -1 \Rightarrow C_v/C_p - C_v \cdot d(pv)/p \cdot dv) = -1 \Rightarrow (1 + (v \cdot dp/p \cdot dv) = 1 - C_p/C_v \Rightarrow v \cdot dp/p \cdot dv = -k \Rightarrow p \cdot v^k = const.$

[†] Equation $p \cdot v^k = const$ is divided term wise by the equation of Clapeyron.

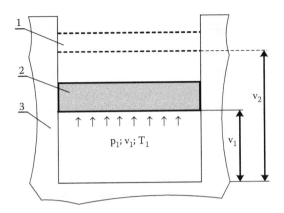

FIGURE 1.18
TDS exchanges only mechanical work with the environment: 1—movable wall; 2—top dead center of the movable wall; and 3—heat insulation.

$$T \cdot v^{k-1} = \text{const} \quad \text{or} \quad T_1 v_1^{k-1} = T_2 v_2^{k-1}, \quad (1.58)$$

where $k = C_p/C_v$ is a constant index of the adiabata.[*]
Both Equations 1.57 and 1.58 produce a relation between the parameters when the TDS is in equilibrium. They prove that TDS specific volume v, on one hand, and pressure p and temperature T, on the other hand, are inversely proportional to each other (i.e., if $v\uparrow$, then $p\downarrow$ and $T\downarrow$, and vice versa). This is illustrated by plots $\bar{p} - \bar{V}$ and $\bar{T} - \bar{s}$ in Figure 1.19.
When a TDS expands from an initial volume v_1 (point "1") to a final volume v_2 (point "2") (i.e., $dv > 0 \Rightarrow v_2 > v_1$), pressure ($p$) and temperature ($T$) drop owing to the inverse proportionality—see Figure 1.19a. The process runs along line "1-2" as shown in Figure 1.19a, b. Pressure (p_2) and temperature (T_2) at the final point "2" are found modifying the expressions 1.57 and 1.58, that is,

$$p_2 = p_1 \left(\frac{v_1}{v_2}\right)^k \quad \text{and} \quad T_2 = T_1 \left(\frac{v_2}{v_1}\right)^{k-1}.$$

The mechanical work $l_{M_{1-2}}$ done on TDS expansion is calculated via the variation of the internal energy ($\delta l_M = -du$), for temperature decrease ($T_1 - T_2$). Hence,

$$l_{M_{1-2}} = -u_{1-2} = -C_v(T_2 - T_1) = C_v(T_1 - T_2). \quad (1.59)$$

when the TDS constricts, the process runs along line ($1 - 2'$), pressure (p) and temperature (T) rise, and the mechanical work is negative, since it is done by external forces (Figure 1.19a,b).
Entropy (s) of the adiabatic process remains constant ($ds = \delta q/T = 0$, and $s_1 = s_2 = \cdots = $ const), since there is no heat exchange with the environment and $\delta q = 0$.

---

[*] $k = 1.67$; 1.4; 1.29, respectively, for one-, two-, and multi-atom gases.

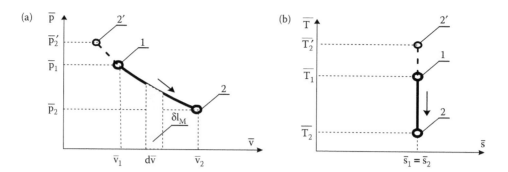

FIGURE 1.19
During an adiabatic process, TDS expansion takes place (positive work is done $l_{1-2} > 0$) at the expense of the internal energy ($u\downarrow$ and $T\downarrow$): (a) mechanical diagram and (b) thermal diagram: 1—initial state of TDS; 2—final state of TDS at expansion; and 2'—final state of TDS at compression.

Hence, adiabatic expansion (1–2) and compression (1 – 2') are mapped by a vertical line in the temperature diagram $T - s$—Figure 1.19.

1.3.5 Generalization of Processes Running in a TDS

The elementary processes described above can be realized under idealized conditions, only. Hence, it is assumed that they can be generalized adopting a power law, similar to the adiabatic one,* and the so-called "polytropic process" is introduced as a generalization, where

$$p \cdot v^n = \text{const} \quad \text{and} \quad T \cdot v^{n-1} = \text{const.} \tag{1.60}$$

Here, $n = -(C - C_p)/(C - C_\vartheta)$ is the polytrope exponent equal to

- $n = 1$—for isothermal process ($T = \text{const}$)
- $n = k$—for adiabatic process ($s = \text{const}$)
- $n = 0$—for isobaric process ($p = \text{const}$)
- $n = \infty$—for isochoric process ($v = \text{const}$)

Figure 1.20 shows four plots of elementary TDPs, denoted by their polytropic index n ($0 \leq n \leq \infty$).

Compare the mechanical and chemical work done for one and the same of TDS compression ratio ($\varepsilon = v_2/v_1$) and one and the same entropy s_2 gained at point 2". Analyzing both types of work, we may conclude that

1. Keeping the condition $\varepsilon = v_2/v_1 = \text{const}$, *the isobaric process* (curve 1–2ᴾ—Figure 1.20a) "extracts" the largest amount of work $l_{M_{1-2}}$ while *the adiabatic process* (curve 1–2ˢ) "extracts" the smallest amount of work. The TDS is not supplied with

---

* Based on the assertion $du/\delta q = \text{cons}$. Using routine assumptions and modifications of the type $du = C_v \cdot dT$ и $\delta q = C \cdot dT$, we find $Cv \cdot dT/C \cdot dT = \text{const} \rightarrow (C-C_v)/R \cdot (d(pv)/p \cdot dv) = 1$.

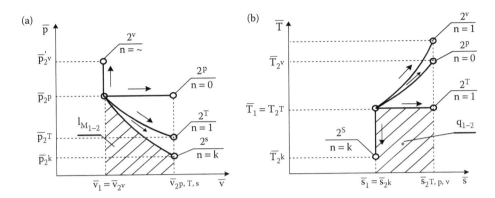

FIGURE 1.20
Generalized diagram of a polytropic process: (a) mechanical work of a TDS and (b) thermal work of a TDS. (Note: The diagram specifies the direction and final equilibrium states at TDS expansion or TDS supply with external heat.) 1—initial state of TDS; 2^T—final state of TDS at isothermal conditions; 2^V—final state of TDS at isochoric conditions; 2^S—final state of TDS at adiabic conditions; and 2^P—final state of TDS at isobaric conditions.

external heat during the operation of the last process, and mechanical work is done at the expense of the internal energy, only.

2. Entropy level $s_2 = const$ is attained under *an isochoric process* by supplying the largest amount of heat (q_{1-2}), and TDS internal energy only increases, without the performance of any mechanical work. The least quantity of heat is supplied to or released from a TDS when *an isothermal process* runs, and the mechanical work is directly converted into heat ($q_{1-2} = l_{M_{1-2}}$) in this case.

Table A.4 of the Appendices specifies formulas valid for the four TDPs and used in engineering calculations to assess the mechanical and thermal work done within a TDS.

1.3.6 Control Questions

1. What is the link between temperature and pressure under isochoric heating?
2. How can one assess the heat supplied during an isochoric process?
3. How can one assess the mechanical work done during an isochoric process?
4. What is the relation between TDS specific volume and temperature during isobaric cooling?
5. External heat q_{1-2} supplied during isobaric heating can be assessed considering: (a) change of the internal energy (u_{1-2}); (b) change of entropy (s_{1-2}); (c) change of enthalpy (h_{1-2}); and (d) mechanical work ($l_{M_{1-2}}$) done.
6. Consider an isothermal process running in a TDS. What is the relation between pressure and volume?
7. Specify what the external heat supplied to a TDS is equal to under an isothermal process.
8. How does an adiabatic expansion of a TDS proceed? At whose expense is the mechanical work $l_{M_{1-2}}$ done in this case?
9. How can a TDS constrict without exchanging heat with the environment (i.e., under adiabatic conditions)?

10. Under which TDP is the largest quantity of mechanical work $l_{M_{1-2}}$ done if the compression ratio is one and the same for all processes?

11. How can one explain this event? Under which process does a TDS gain the largest quantity of heat if entropy attained at the TDS final state is one and the same regardless of the process running?

1.4 Energy Conversion Machines and Their Thermodynamic Cycles: A Standard for Energy Efficiency—Carnot Cycle

1.4.1 Energy Conversion Machines: Basic Types

Energy conversion (power) machines are aggregates of material bodies linked to each other according to a specific scheme and with a task to convert one form of energy to another. For instance, engines (thermal or electric) convert energy (chemical or electrical) to mechanical. Similar is the task of Seebeck converters and fuel cells or that of magneto-hydrodynamic generators, the latter converting mechanical motion of cold or hot plasma to electrical energy. Those power machines are known as *generators* or *cogenerators*. Note that the term *cogenerator* is popular in areas where, besides the basic energy (electric power, for instance), by-products treated as waste in traditional generators (heat, for instance) are also made use of.

Other power machines change only the characteristics of a certain form of energy, keeping its amount. These are mechanical gears, reducers, multiplicators, and friction devices where only kinematic and power parameters of the mechanical energy (rotational, reciprocal-translational, or reciprocal-oscillation) vary. On the other hand, electric transformers, including magnetic-induction or electronic ones, do not convert the electrical energy to another energy form but change its characteristics (voltage, frequency, or power), only. Heat transformers, including heat pumps and refrigerators, are of the same type and increase or decrease temperature, reminding the thermal energy form as an intake and exhaust of the devices.

Having in mind the wide variety of power machines, it is convenient to classify them as follows:

- *Energy generators* (devices converting primary energy to "useful" energy which then is carried by various carriers—electric current, steam, hot water, conditioned air, etc.).

- *Energy transformers* (devices changing the parameters of energy, keeping its form).

- *Operating machines* (aggregated devices producing "useful" energy in the form of work or heat). They can be classified as

  - *Engines*—devices performing mechanical work to facilitate labor and living (employed in industry, households, and transport)

  - *Appliances*—devices used in industry or households and producing energy in the form of electromagnetic radiation or convective fluxes (ovens, cooking stoves, electric heaters, fryers, toasters, etc.)

  - *Converters*—devices producing acoustic energy or radiation (radiators, radiating bands, light sources, audiovisual, and communication devices)

The classification proposed is conditional and intended to facilitate communication between specialists.

1.4.2 Thermodynamic Requirements to the Power Machines

Despite the wide variety, commercial power machines[*] should satisfy three general requirements:

1. They should guarantee cyclic operation expressed in multiple repeatedness of the energy exchange
2. The capacity of the primary energy sources should be completely used, that is, energy efficiency of the devices should be maximal
3. Any device should be assembled using recyclable and environmentally friendly materials and components

As outlined in Section 1.2.2, to design a machine with cyclic operation, a certain combination of TDPs should run along a closed loop, multiply passing through its starting point—see Figure 1.21a. This, however, cannot be realized without successive contacts between the TDS and hot and cold energy sources in order to maintain the natural course of heat exchange.

How can a closed cycle be organized? This is formally illustrated by the mechanical diagram (p–v). Figure 1.21a shows a two-phase "closed" cycle as an example. Formally, phases are denoted as

- Phase "1-A-2"
- Phase "1-B-2"

Their characteristic feature is the different amount of mechanical work done (the area under line "1-A-2" is larger than that under line "1-B-2"). Hence, it follows that the *sequence of phase "coverage"* within a closed cycle is decisive for the balance of the general mechanical work done during the full cycle. According to that sequence (the order of occupying the initial and final states), a TDS may do work "against" the environment, that is, it may "gain" profit, or in contrast, the environment may constrict the TDS doing negative mechanical work per cycle, that is, an entire revolution of energy conversion may be completed. When a TDP runs along the sequence "1-A-2-B-1" (Figure 1.21a), the total work is positive since

$$L_{II} = l_{1-A-2} - l_{2-B-1} > 0, \tag{1.61}$$

and $\left| l_{1-A-2} \right| > \left| l_{2-B-1} \right|$. Then, the TDS expands and performs useful mechanical work. The travel over the thermodynamic states is directed as ("1-A-2-B-1"), and travel (cycle) direction is clockwise. It is assumed as "straight" direction, and the cycles are called *clockwise cycles*. Clockwise cycles present a theoretical basis of the design of a number of heat engines.[†]

---

[*] For widespread use.

[†] The first operating heat engine was designed by Denis Papin (1647–1714) and consisted of a piston performing vertical translation and a lever—Figure 1.11. Later Papin's engine was improved by T. Newcomen (1664–1729) and supplied with a pump. It was used in mining in England in the 18th century.

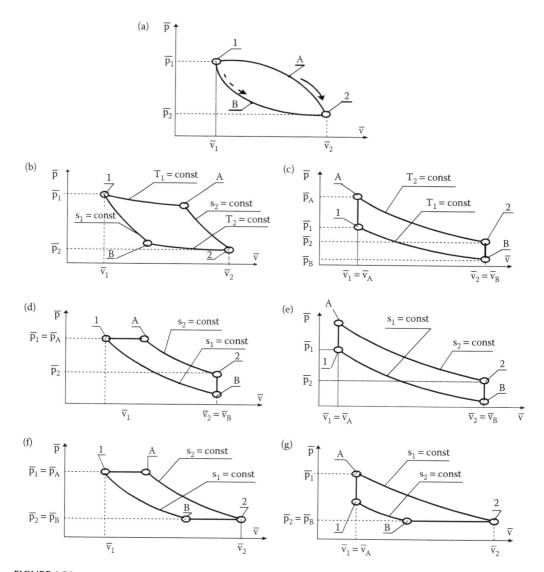

FIGURE 1.21
TDPs of TDS variation along a closed loop: (a) general scheme of a "closed" TDS; (b) Carnot cycle; (c) Stirling cycle; (d) Diesel cycle; (e) Otto cycle; (f) Brayton cycle; and (g) Humphrey cycle.

The inventor of the first heat engine, consistent from an engineering point of view, was the Scottish engineer James Watt (1736–1819). In 1784, he patented an industrial version of the steam engine whose scheme is shown in Figure 1.22. Quite soon Watt's engine was applied not only in industry but also in transport, and it became a basis of the industrial revolution that conquered Europe and North America.

Still at the time of the first heat engines,* people realized that TDPs running in opposite sequence, for instance "1-B-2-A-1," could be realized in devices where the TDS state param-

---

* The first refrigerator was constructed by William Cullen in 1784. Oliver Evans designed the first refrigerator performing phase transformation in the refrigerating aggregate in 1805, and the first refrigerating wagon was manufactured in 1857.

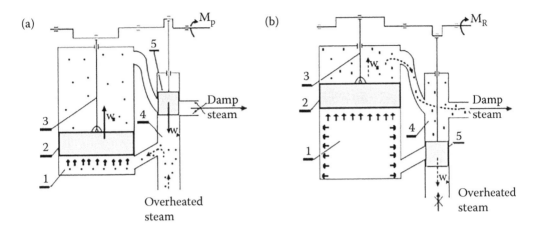

FIGURE 1.22
Scheme of the steam engine of J. Watt operating with superheated steam: (a) lower dead center and (b) upper dead center. 1—power cylinder; 2—piston of slider-crank mechanism; 3—crank; 4—steam distributor; and 5—distributing pistons.

eters, for example, temperature (T) or pressure (p), could be purposefully varied. Indeed, when a closed TDP runs counterclockwise, the total work of the cycle is negative

$$L_ц = l_{1-B-2} - l_{2-A-1} = l_{разш} - l_{комп} < 0. \tag{1.62}$$

Here, TDS expands along the line "1-B-2" and performs positive work l_{1-B-2}. After occupying the final state "2," compression takes place along the line "2-A-1," where the mechanical work done is negative (i.e., $l_{2-A-1} < 0$) but its absolute value is larger than that done on TDS expansion. Hence, the balance of the mechanical energy is negative, and to function, those cycles need import of work.

Thermodynamic cycles where a TDS "hires" work from the surroundings to change its state parameters are called *counterclockwise cycles*. Technically, that "consumption" or import of mechanical work is realized using natural forces (gravity, electromagnetic forces, etc.) or employing external engines. Counterclockwise cycles are the basis of the design of two vast groups of power machines—refrigerators and heat pumps, which are widely applied in the food, wine, and tobacco industries and in the air-conditioning of buildings, restaurants, and vehicles. However, an entire TDP runs along a closed loop (regardless of the travel direction) combining various elementary TDPs of the type discussed in Section 1.3 and illustrated in detail in Figure 1.21b–g. Note that the figures present closed cyclic processes that are technically attainable, only, and the construction of a number of power machines is based upon them.

The combinations of TDPs are not restricted to those shown in Figure 1.21, only, but each one can be improved and modified. For instance, the combination shown in Figure 1.21e consisting of two adiabatic (s_1 = const, s_2 = const) and two isochoric (1-A and 2-B) processes is classical and is known as the Otto cycle (applied in carburetor thermal motors). However, it was modified by Adkinson in 1882 by combining six (instead of four) elementary processes within a cycle. An extra isochoric process and an extra isobaric process were added to the processes shown in Figure 1.21e. In its turn, Adkinson cycle was improved by R. Miller in 1947. He introduced simple mechanisms to change the duration of filling and cleaning of the combustion chamber. Thus, 20% lower fuel consumption and

70% lower toxicity of waste gases were attained. Similar examples of improving thermo-dynamic cycles and power machines will be discussed in Chapter 2.

The cyclic mode of operation of the closed TDPs does not automatically provide solution to the problem of energy efficiency. The first heat engines realized cyclic continuous operation but the energy efficiency was too low and not exceeding 2%–3%. Hence, large portion of the primary energy (released during oxidation of biomass or ground coal), was converted into waste heat instead of useful mechanical energy.[*]

As for thermodynamics, the young French researcher *Nicolas Léonard Sadi Carnot* was the first to analyze and solve the problem of energy efficiency in his work of genius "Reflections on the Motive Power of Fire" published in 1824.[†] Carnot studied the already designed heat (steam) engines and proposed a structure and principles of operation of an ideal thermal device. To this date, such a device has not been constructed but the TDP proposed was named "Carnot cycle" and used as an "energy standard" by a number of designers. It is still in use to predict the maximal power effect of a certain technology. Due to its essential methodological importance in assessing efficiency of energy conversion, we will discuss it in detail in what follows, considering two types of power machines—heat engines and heat transformers.

1.4.3 Carnot Thermodynamic Cycle Used as an Energy Standard

Nicolas Carnot, speculating on the effective use of primary energy, supposed that a heat engine should operate within a clockwise closed thermodynamic cycle consisting of two isothermal and two adiabatic processes, only (see Figure 1.23e):

1. Isothermal expansion (point 1 → point A)
2. Adiabatic expansion (point A → point 2)
3. Isothermal compression (point 2 → point B)
4. Adiabatic compression (point B → point 1)

As proposed by Carnot, an isothermal process is the first step of an ideal efficient cycle, since the total external heat q_1 (temperature T_1) is consumed to expand the system and do mechanical work (see Section 1.3.3). At this stage (point 1–point A), the hot source (HS) supplies heat q_1 to the TDS—Figure 1.23a. Keeping the energy balance of the isothermal process (Equation 1.52), the form of the mechanical work is

$$l_{M_{1-A}} = RT_1 \ln \frac{v_A}{v_1} = R_\mu T_1 * \ln \frac{p_1}{p_A},\qquad(1.63)$$

since $T_1 = $ const and $du = 0$. The necessary amount of heat supplied is $q_1 = T_1 (s_A - s_1)$.

Adiabatic expansion takes place during the second step, since the TDS neither gains nor loses heat, and the total mechanical work is done at the expense of TDS internal energy, only, and respectively—at the expense of decrease in temperature from T_1 to T_2. Note also that due to the positive work done, the TDS expands to a specific volume $v = v_2$—see Figure 1.23b. The second hierarchical position of this step is not randomly chosen, since the TDS has kept its internal energy intact during the first step. Work done $l_{M_{A-2}}$ is assessed applying the law of energy conversion specified by relation 1.31, where $l_{M_{A-2}} = -u_{A-2} = C_v(T_A - T_2)$.

---

[*] Efficiency of modern internal combustion engines amounts to 28%–30%.
[†] Reflexions sur la puissance motrice du feu et sur les machines propres a developer cett puissance.

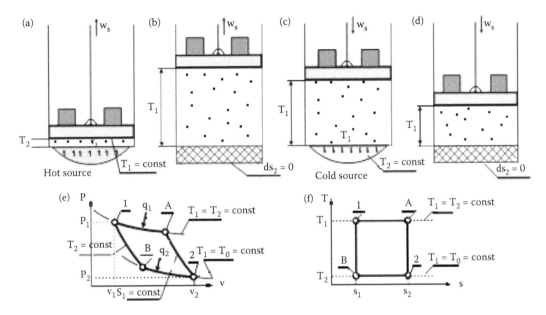

FIGURE 1.23
Carnot's ideal thermal machine: (a–d)—cycle phases and (e,f)—mechanical and thermal diagram: 1-A—isothermal expansion (point 1 → point A); A-2—adiabatic expansion (point A → point 2); 2-B—isothermal compression (point 2 → point B); and B-1—adiabatic compression (point B → point 1).

During the third step (Figure 1.23b), TDS isothermal compression takes place via connection of the system to a cold source. Heat q_2 is effectively released maintaining the temperature T_2, and the TDS intensively cools down. As a result of cooling and compression of the control volume, the pressure rises from p_2 to p_B, and the TDS reaches volume (v_B), while entropy drops from s_2 to s_B.

Heat q_2 released to the cold source amounts to

$$q_2 = -T_2(s_2 - s_B) \rightarrow (\text{see Figure 1.23f}),\tag{1.64}$$

and the work for compression of the working volume is done at the expense of q_2:

$$l_{M2-B} = -R_\mu \cdot T_2 \ln \frac{v_2}{v_B},\tag{1.65}$$

since $T_2 = \text{const}$ and $du = 0$ (see Figure 1.23e).

Adiabatic compression of the TDS takes place during the fourth (last) step of the Carnot cycle (Figure 1.23d), where pressure rises from p_B to p_1, and temperature increases from T_1 to T_1. Without heat release to the surroundings ($ds = 0$), the external mechanical work done for TDS constriction l_{MB-1} increases the amount of internal energy ($u_B \Rightarrow u_1$) and the temperature rises from T_2 to T_1, so that

$$u_{B-1} = C_v(T_1 - T_B) = C_v(T_1 - T_2) = -l_{MB-1},\tag{1.66}$$

that is, the work is negative.

Cycle total work per TDS unit mass is the sum of the four works

$$L_C = l_{M1-A} + l_{MA-2} - l_{M2-B} - l_{MB-1} = q_1 + C_v(T_1 - T_2) - q_2 - C_v(T_1 - T_2) = q_1 - q_2. \quad (1.67)$$

Energy efficiency of the *clockwise Carnot cycle* is estimated by the ratio between the useful work (L_C) and the supplied external heat q_1, known as "coefficient of performance":

$$\beta = \frac{\text{mechanical work done during a single cycle}}{\text{supplied external heat}}$$

$$= \frac{L_C}{q_1} = \frac{q_1 - q_2}{q_1} = 1 - \frac{q_2}{q_1}. \quad (1.68)$$

The thermal diagram T–s (Figure 1.23f) shows that heat q_1 supplied by the HS is

$$q_1 = T_1(s_A - s_1), \quad (1.69)$$

and heat q_2 released to the cold source is

$$q_2 = T_2(s_2 - s_B). \quad (1.70)$$

Since $(s_A - s_1) = (s_2 - s_B)$, the coefficient of performance β takes the form

$$\beta_{Carnot}^{Clockwise} = 1 - \frac{T_2}{T_1}. \quad (1.71)$$

The above relation shows that the energy efficiency of *Carnot's ideal heat engine* is specified by the temperatures of the CSs and HSs—T_1 and T_2. Efficiency increases when the temperature ratio T_2/T_1 decreases for ($T_2\downarrow$) and ($T_1\uparrow$).

When a power machine operates under counterclockwise *Carnot cycle* and the positive effect consists in that the cold source releases heat q_2 (as is in a refrigerator), efficiency is estimated by the ratio

$$COP = \beta_{refr}^{Carnot} = \frac{q_2}{|L_{II}|} = \frac{q_2}{q_1 - q_2} = \frac{1}{T_1/T_1T_2T_2 - 1} \quad (1.72)$$

called "coefficient of performance."

When a power machine operates as a heat pump and the positive effect consists in supplying the HS with heat q_1, efficiency is estimated by a "conversion coefficient" of the form

$$\eta_{Thermopump}^{Carnot} = \frac{q_1}{|L_C|} = \frac{1}{1 - T_2/T_1} > 1 \rightarrow T_1 > T_2. \quad (1.73)$$

Estimations show that the energy efficiency of an ideal refrigerator and a heat pump depends on the ratio between the temperatures of the cold and warm sources—(T_2)

and (T_1), only. Efficiency of real refrigerators will be discussed in the second part of the book.

1.4.4 Control Questions

1. To which class of power machines do energy converters belong?
2. To which class of power machines do parametric energy converters belong?
3. What does energy cogenerator mean?
4. Do operating machines comprise executing mechanisms?
5. Which are the basic thermodynamic requirements to power machines?
6. How can one attain cyclicity of the operation of a power machine?
7. If the total work of a TDS is negative, how could one physically provide the necessary work for TDS constriction?
8. How many are the ways of "constructing" a closed thermodynamic cycle? Specify the most popular thermodynamic cycles.
9. What is the application of the clockwise thermodynamic cycles?
10. How could one distinguish the clockwise from the counterclockwise cycle?
11. Who is the founder of the theory of energy efficiency of thermodynamic cycles? Are you familiar with details of his life?
12. Who built the first heat engine of industrial importance? How was this engine built?
13. What was the energy efficiency of the first thermal engines? What is that of modern engines?
14. Which are the TDPs that constitute the thermodynamic cycle of an ideal heat engine? Why were those processes chosen?
15. How can one estimate the energy efficiency of a heat engine?
16. How can one increase the energy efficiency of an ideal heat engine?
17. How can one estimate the energy efficiency of heat transformers—refrigerators and heat pumps?
18. Can the conversion coefficient of a heat pump be less than 1?
19. Why are energy users interested in the technological application of heat pumps?

1.5 Specific Features of Real Energy Conversion

1.5.1 General

So far, the main theoretical concepts of analysis and explanation of thermal-energy processes, technologies, and principles of heat engine operations were discussed. Moreover, idealized physical restrictions were imposed on the working fluid, concerning in particular interatomic–intermolecular forces and viscosity which affect gas motion in a TDS. For instance, a direct consequence of disregarding *the Van der Waals forces* and molecular and atomic volumes is that phase transformations, throttle-valve or turbulent effects are not considered in the study of processes that run in ideal power machines. Such a stipulation, however, is admissible in preparing the introduction of academic courses and technical books.

Yet, to gain profound knowledge on real energy conversion occurring in a TDS, we shall additionally consider the following physical phenomena that the above idealizations disregard:

- Phase transformations of heat transferring fluids in a TDS
- Cavitation
- Throttle-valve effects
- Vortex effects

Their consideration leaves a specific mark on the processes running in a TDS, on the efficiency of thermal-energy technologies and on the structure of power machines. Last, knowing their effect on TDS performance is important not only for machine design but also for machine exploitation.

1.5.2 Phase Transformations of a TDS Working Body

The working ideal gas (fluid) of all idealized TDS discussed so far is assumed to have only one aggregate state—vapor state. Ideal gas, as a working body where various processes run, is able to store and release heat. Its transformation into liquid or solid is a priori excluded. Real working bodies, however, may adopt three aggregate states depending on the degree of current integrity of their molecules and atoms. Transition from one into another state depends on their capability to accumulate internal energy u or, respectively, to change their temperature.

Table 1.3 specifies temperature at which the four most popular substances change their aggregate state at normal pressure.

Data prove that at room temperature of 20°C (293.16 K) water is liquid, iron is solid, and hydrogen and oxygen are gases. Yet, below 14 K all of them are solids.

Although that both transitions (solid/liquid and liquid/vapor) are of interest for power engineering, the liquid/gas transition is exploited on a wider scale in modern energy-generating (power) technologies and is of special scientific and public interest. Hence, we shall focus on it. Note that it is also popular as "evaporation/liquefaction" or "boiling/condensation." In particular, since water is one of the most common working fluid, we shall treat in detail its liquid/vapor and inverse transformation in what follows (see Figure 1.24). Similar mechanisms of transformation can also be uncovered for other substances used as TDS working fluids.

TABLE 1.3

Temperatures of Phase Exchange

$p = 10^5$ Pa	Solid/Liquid $T_{melting}$, K	Liquid/Vapor $T_{boiling}$, K
H_2O	273.16	373.10
H_2	14.01	20.28
O_2	50.3	90.18
Fe	1808.0	3023.0

The process of "liquid/gas" transformation of a substance is demonstrated here in a cylinder with a movable wall and filled with fixed amount of water, while pressure is constant ($p_1 = G/A_0$ = const—see Figure 1.24a–d and $p_2 = 2G/A_0$ = const—see Figure 1.24e–g.

Heat q is supplied through the vessel bottom and phase transition occurs during the four stages along the loop "A-B-C-D-E":

- Isochoric heating ("A–B") to pressure p_2, equal to the external pressure.
- Isobaric heating ("B–C") to T_S (water saturation temperature). Here, internal energy becomes sufficient to initiate phase transition of water (see Figure 1.24a). Initially, the process starts as evaporation and before long converts into boiling.
- Isobaric boiling ("C–D"). The whole amount of intaking heat q is spent to break the intermolecular bonds. Hence, the temperature of the working fluid are kept constant ($T = T_S$ = const).

Evaporation proceeds until all molecules moving freely leave the liquid as vapor (Figure 1.25b). That process grows more and more intensive and proceeds as boiling. At the same time, TDS specific volume grows ($v = v_K$), owing to the increase of the distance between the free molecules within the vapor "pocket" formed under the movable wall.

The fourth phase starts after evaporation of the whole amount of liquid. It is isobaric superheating ("D–E")—Figure 1.24d. Here, the whole amount of heat supplied to the TDS is spent to increase enthalpy h and temperature T of the free molecules. Intermolecular distances and vapor specific volume also grow, and vapor starts behaving as an "ideal gas."

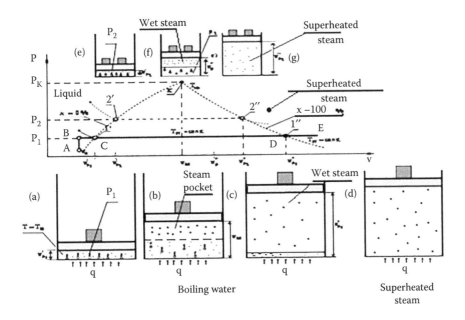

FIGURE 1.24

Illustration in the p–v diagram of vapor generation (at 273.16 K and v = 0.001 m³/kg): (a) boiling initiation ($p = p_1$, $T = T_S$; x = 100%); (b) vapor fraction of two-phase mixture ($p = p_1$; $T = T_S$; 0 < x < 100%); (c) dry vapor ($p = p_1$ $T = T_S$; x = 100%); (d) dry vapor ($T > T_S$); and (e–g) phase transition at increased pressure ($p = p_2$; $x = v_x/v'$)—vapor share of the two-phase component.

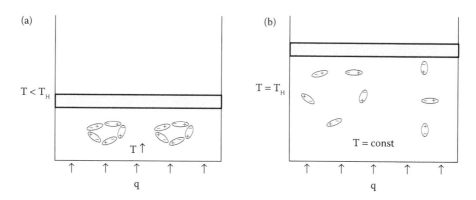

FIGURE 1.25
Water molecules are released due to Van der Waals forces, after attaining u_s and T_s. Release is initiated at point C (see Figure 1.24a): (a) water molecules with strong polarity form coalitions ($T < T_s$) and (b) intermolecular bonds break with the increase of $T \Rightarrow T = T_s = $ const, but still remain active within the liquid.

When phase transformation of water takes place at higher pressure—p_2 for instance (Figure 1.24e–g), the process comprises the same phases. Yet, it has some specific features:

1. Boiling initiation retards ($v'_{p2} > v'_{p1}$)
2. Vapor specific volume (see Figure 1.24g) is smaller ($v''_{p2} < v''_{p1}$)
3. The area of "two-phase vapor/water mix" is narrower ($2' - 2' < 1' - 1'$)

(see the illustration in Figure 1.26—plots p–v and T–s).
The figure shows three outlined regions of water aggregate states:

- Liquid phase
- Two-phase region (boiling water and wet vapor)
- Gaseous phase (dry or superheated vapor)

The regions are bordered by lines 1'-2'-3'-⋯-K (boiling start) and K-3''-2''-1'' (boiling end and superheating start). "Aggregate state diagrams" are plotted visualizing phase transformations. It is found that when energy transformations take place in the transition

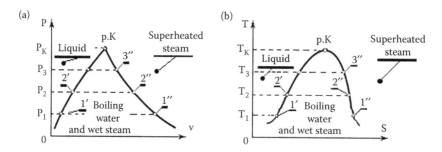

FIGURE 1.26
Connected points 1', 2', 3', …, K, 3'', 2'', 1'' outline three phase regions: liquid, boiling liquid with wet vapor, and dry vapor. Point "K" is called "critical point" where two-phase transition runs (from liquid directly into superheated phase). (a) P–v diagram and (b) T–S diagram.

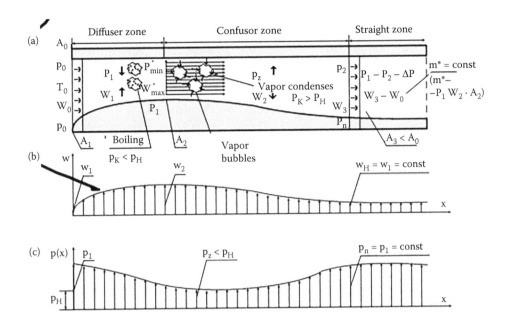

FIGURE 1.27

Open TDS with variable stereometry (geometry): "nozzle" configuration. (a) Longitudinal section of a channel with variable cross section; (b) velocity distribution; and (c) pressure distribution.

"liquid–vapor" regions, paths of the TDPs significantly differ from those running in the dry vapor region.

Consider for instance the mechanical diagram p–v. As seen, the Carnot cycle differs from its traditional shape in the case of *an ideal gas operating as a working fluid* (compare Figure 1.27a to 1.23e).

This is so, since both isothermal processes run here under current boiling/condensation of the working fluid, and intermolecular forces dominate (heat breaks or creates intermolecular bonds). Hence, plots differ significantly from those for an "ideal gas." Some of the relations used to find the state parameters of the working fluid within the three regions are briefly specified in Table A.6 of the Appendices. The state diagrams of other working fluids such as refrigerant R-134-a, R-404A, R-410A, and R-507A often used in thermotechnic practice are also given in Tables A.7 and A.8 of the Appendices.

1.5.3　Cavitation

The specific phenomena in a TDS—cavitation and phase transformation, have a common physical basis—operation of intermolecular forces. Moreover, cavitation can be considered as a TDP consisting of two successive changes of the aggregate state of the working fluid. It runs in an insulated but open TDS. The two phases are

- Boiling ("liquid ⇒ vapor" transition)
- Condensation ("vapor ⇒ liquid transition")

This TDP takes place during a short time interval within a single microfluidic volume of a working medium. It is observed in a TDS whose controlled boundaries have a

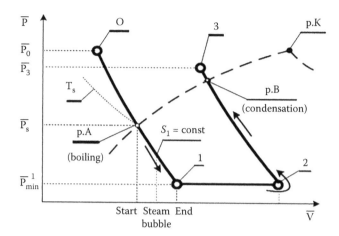

FIGURE 1.28
Thermodynamic interpretation of cavitation—consisting of two successive phase transitions: boiling at point A and condensation at point B.

variable geometry (see Figure 1.28). Consider an open but insulated TDS, that is, a channel with variable cross section $A_0 = f(x)$ where a fluid flows (water, chlorofluorocarbons, ammonium, etc.). Then, three specific zones are outlined resulting from channel specific geometry:

- Confusor zone with decreasing cross section—constricting zone ($dA_0 < 0$ or $A_0\downarrow$)
- Diffuser zone with increasing cross section—expanding zone ($dA_0 < 0$ or $A_0\uparrow$)
- Straight zone with constant cross section ($dA_0 = 0$ or $A_0 = $ const)

To assess the initiation and evolution of cavitation, one should account for the parameter of working fluid change within those zones. Assume that the working fluid is incompressible* ($\rho = $ const), and the process is adiabatic ($\delta q = C_v \cdot dT = 0 \Rightarrow T = $ const). Then, both forms of the law of conservation will read

- *Mass conservation* ($m^* = $ const):

$$A_1 \cdot w_1 = A_2 \cdot w_2 = A_3 \cdot w_3 \quad \text{per 1 kg} \tag{1.74}$$

and

$$w_2 = w_1 \frac{A_1}{A_2} \quad \text{and} \quad w_3 = w_1 \frac{A_3}{A_2}. \tag{1.75}$$

- *Energy conversions:* using Equation 1.36 for ($\delta q = 0, dz = 0$) and $n = 1, 2, 3$, we obtain Euler equation of an ideal fluid. It has the following form for the three zones:

$$(vp)_1 + \frac{w_1^2}{2} = (vp)_2 + \frac{w_2^2}{2} = (vp)_3 + \frac{w_3^2}{2} \tag{1.76}$$

---

* Water is practically incompressible for velocity lower than 100 m/s.

TABLE 1.4

Change of the Velocity and the Pressure in Those Parts of the Nozzle

Area	Cross Section	Area Velocity	Area Pressure
Confusor zone	$A_1\downarrow$	$w_1\uparrow$	$p_1\downarrow$
Diffuser zone	$A_2\uparrow$	$w_2\downarrow$	$p_2\uparrow$
Straight	$A_3 = A_0 = \text{const}$	$w_3 = w_0 = \text{const}$	$p_3 = p_0 = \Delta p_{TP} = \text{const}$

or

$$p_2 = p_1 - 0.5\rho(w_2^2 - w_1^2) \quad \text{and} \quad p_3 = p_1 - 0.5\rho(w_3^2 - w_1^2). \tag{1.77}$$

Noting the varying area A_0 of the channel cross section, one can follow the evolution of the TDS parameters (see Table 1.4 and Figure 1.28 and the above equations):

- *Confusor zone:* velocity increases and pressure drops
- *Diffuser zone:* the opposite tendency is observed—velocity decreases and pressure rises
- *Straight zone:* current and initial velocities equalize ($w_3 = w_0$), and pressure also approaches its initial value

(friction losses Δp_{TP} are subtracted[*]).

The variation of the TDS parameters thus outlined proves that pressure p_1 steadily drops in the confusor zone. If it attains the saturation value p_S, boiling of the working fluid will start and formation of vapor bubbles in the liquid will be observed (i.e., vapor bubble generation—see Figure 1.27a).

Since p_1 drops along the flow, the minimal value p_{min} will be attained at the narrowest cross section of the channel, and boiling will be most intensive there.

Vapor bubbles move together with the liquid, and they leave *the first zone* to enter *the diffuser zone* where velocity drops and pressure rises. When pressure reaches a value such that $p_1(x) > p_S$, vapor condenses and the working fluid is transformed into liquid, while vapor bubbles collapse. After vapor condensation, the liquid rushes into the freed vapor micro volumes, and intensive energy interactions can take place in the locations of vapor bubble collapse, namely:

- Shock wave formation
- Generation of pulsations with significant amplitude acting on the border wall (the TDS control surface)
- Generation of wide spectra noises, including ultrasonic noises
- Occurrence of fast cavitation fracture of the wall[†]
- Occurrence of luminescent effects (sono luminescence)
- Intensive mixing and turbulence of the working fluid

---

[*] $\Delta p_H = \lambda_p(I_p / d_p).0, 5\ \rho w_p^2$—linear losses due to friction.
[†] Materials, most resistant to cavitation, are cast steel St45, aluminum bronze, nickel steel (2% Ni), cast molybdenum steel, tin bronze, etc.

- Development of chemical reactions (liquid dissociation—water decomposes into O_2 and $2H_2$)
- Dissipation of the hydraulic mechanical energy (it is converted into heat)

Cavitation takes place at the boundary between the confusor zone (*boiling*) and the diffuser zone (*condensation*). As any TDP, it can be presented by a diagram of the working fluid state $(p-v)$—see Figure 1.28. Cavitation commences at point 0, which belongs to a specific microfluid section of the channel confusor zone. Due to the increased local velocity, pressure p_1 in the confusor zone drops $(w\uparrow, p\downarrow)$—Figure 1.27. At point A pressure p_1 equalizes with p_s, *and liquid boiling starts resulting in the formation of vapor bubbles.*

When pressure reaches value, its adiabatic increase starts (proceeding from point 2 to point 3) due to the change of the control area geometry. At point B (close to point A), pressure again exceeds p_s, and vapor sharply condenses. This process is continuous; since the system is open and continuous exchange of working fluid (continuous cycle) takes place in the regions of phase transformations. The cavitation effects (incl. turbulent mixing, noise generation, and hydroenergy dissipation) are made use of in liquid mixers, ultrasonic cleaners, cavitation heat generators, and other equipment.

1.5.4 Joule–Thompson Effect in Throttle-Valve Devices

Joule–Thompson effect is the third distinguished manifestation of the operation of intermolecular and interatomic forces in real gases (vapor phase) used as working bodies in open noninsulated thermodynamical systems. It is due to the possibility by varying the geometry of the control area to affect the relation between the components of the total internal energy (enthalpy).

Equations that follow from the laws of mass and energy conversion are valid for real gases. They have the form[*]

$$\delta q = dh - vdp = dh + \frac{dw^2}{2}. \tag{1.78}$$

If there is no heat exchange with the environment $(\delta q = 0)$ and no change of the kinetic energy of the working fluid $(0.5d(w^2) = 0)$, the process runs adiabatically, similar to cavitation, that is,

$$\delta h = 0. \tag{1.79}$$

For i different cross sections of the control surface, the above equality reads

$$u_0 + p_0 v_0 = u_2 + p_2 v_2 = \cdots = u_i + p_i v_i = \ldots, i = 0,1,2,\ldots. \tag{1.80}$$

Regarding open power machines, one needs to perform frequent structural modifications of the control area geometry (constriction and/or expansion). For instance, waste gases of the heat engines outflow through valves (Figure 1.29a), pressure in refrigerators

---

[*] $-vdp = 0{,}5d(w^2)$—the relation follows from Bernoulli equation for real fluid. Note that enthalpy **h** is a sum of the internal kinetic energy (u_i) and the potential energy (pv).

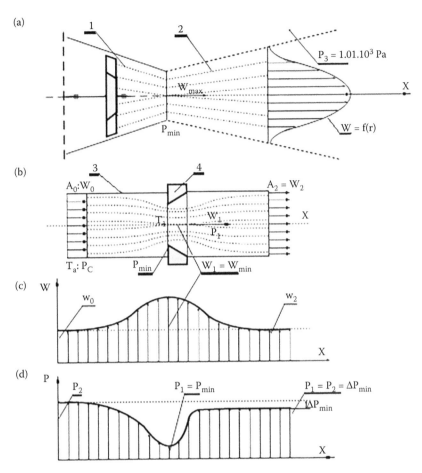

FIGURE 1.29
Sudden constriction of the control surface: 1—exhausting nozzle of a jet engine; 2—free jet; 3—tube; and 4—throttle-valve. (Velocity rises up to w_{max} (Figure 1.29c), and pressure drops to p_{min} (Figure 1.30d).) (a) Flow geometry behind jet engine; (b) flow geometry behind free jet; (c) velocity distribution; and (d) pressure distribution.

is reduced by throttle-valves (Figure 1.29b), etc. Other devices, such as measurers of gas consumption (outflow measurement devices equipped with throttle-valves) and regulators with iris valves are also used in power machines. Similar to Figure 1.27, flux patterns of those devices outline two areas:

- Area of fluid compression corresponding to flow through nozzle 1 and before passing through throttle-valve 4
- Area of fluid expansion (free outflow—after passing through the throttle-valve)

As additional illustrations of valve flux patterns, Figure 1.29c,d shows the distribution of the state parameters—velocity and pressure. Qualitative similarity with Figure 1.27b,c is evident not only in the flux line geometry but also in the distribution of the state parameters. The difference lies in the existence of larger velocity and pressure gradients along motion direction.

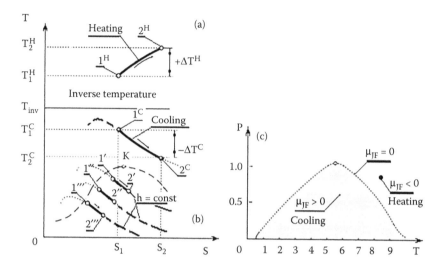

FIGURE 1.30

Adiabatic expansion in the T–s phase diagram: (a) heating of the working body $(T_{2S} > T_{inv})$; (b) cooling of the working body $(T_{2S} < T_{inv})$; and (c) Joule–Thompson coefficient $\mu_{JF} = v/C_p (\alpha T - 1)K/Pa$. The body cools down if $\mu_{JF} > 0$.

When the working fluid is vapor, cavitation in the area of the largest flow constriction does not occur, and since entropy h is kept constant it seems that temperature would stay constant, too (T_i = const). This erroneous conclusion was proven incorrect by Joule and Thompson in 1852 who found that the temperature of the working fluid would decrease or increase (i.e., T ≠ const) depending on the initial conditions—pressure (p_0) and temperature (T_0). Their tribute was the proof of the existence of a limit temperature (the so-called inverse temperature—T_{inv}),* above which the temperature of an outflowing gas increases. Hence, this effect is known as the "Joule–Thompson effect."

The process of heating/cooling during gas adiabatic outflow is illustrated in Figure 1.30 within two temperature ranges of the working fluid:

- Above the "inverse temperature"—"T_{inv}" when the temperature of the working fluid increases—Figure 1.30a

- Below the "inverse temperature" when that temperature decreases—Figure 1.30b

As stated at the beginning of Section 1.5, the Joule–Thompson effect is due to the intermolecular forces (Van der Waals forces) acting in real gases. It results from abrupt compression and subsequent expansion of TDS control volume, and flow geometry and kinematics adapt to those volume changes.

Although this entropy remains constant, its potential component (pv) varies under jump-wise change of pressure at the narrowest cross section. Thus, subsequent change of

---

* T_{inv} depends on the chemical composition. For helium and hydrogen, $T_{inv} = 51$ K (–222°C) and their temperature increases under normal adiabatic outflow. For nitrogen and oxygen, we have $T_{inv} = 621$ K and $T_{inv} = 764$ K, respectively, and they cool down under normal adiabatic outflow. The behavior of the atmospheric air composed of these two elements is similar.

entropy kinetic component and body temperature also takes place. Yet, opposite intermolecular mechanisms operate in the two ranges (above and below T_{inv}).

At low temperature ($T < T_{inv}$), the specific volume of the working fluid increases ($v\uparrow$) together with its expansion, and the distance between molecules grows. Van der Waals forces also increase (similar to the spring force depending on spring deformation degree, they depend on the distance between the molecules). Thus, gas potential energy increases. Hence, to conserve gas total energy (entropy), the internal kinetic energy decreases yielding decrease of temperature (T) (see Figure 1.30b), and processes "1^C–2^C" run in the gas phase region while processes: (1'–2'); (1"–2"); (1'''–2''') run in the two-phase region.

At high temperature ($T > T_{inv}$), despite gas expansion, molecules coalesce much more intensively due to their great internal kinetic energy. Besides, they collide decreasing their potential energy, which in turn results in the increase of the internal kinetic energy and gas temperature. This effect is proved in Figure 1.30a by the course of the process "1H–2H."

Under normal conditions, TDS working fluid such as air and superheated steam operate below the inverse temperature, where the first mechanisms of increase of potential energy prevail. In outflow through nozzles or valves, those gases cool down (air—at $T_{inv} \approx 100$ K, and steam—at $T_{inv} \approx 3000$ K).

The Joule–Thompson effect occurs spontaneously in the components of power facilities. However, it can be used under control in technological devices producing large amounts of compressed air.

1.5.5 Vortex Effects: A Fluid Running into a Chamber

Consider again an open insulated TDS whose control surface is the solid wall of a cylindrical chamber (item 3 in Figure 1.31). The working body is gas (pure atmospheric air) with

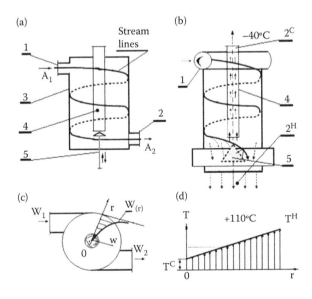

FIGURE 1.31
Vortex effects: (a) cylindrical chamber; (b) Ranque vortex tube; (c) velocity distribution; and (d) temperature distribution: 1—air intake; 2—exhaust; 3—wall; 4—pipe of cold air outflow; and 5—regulating flap.

initial parameters (T_1; p_1; w_1) flowing through a tangential aperture (item 1) with cross section A_1.

All stream lines at the cylinder entrance are parallel and have one and the same mean fluid velocity w_1. When fluid particles enter the cylinder, they start rotating about the cylinder centerline, and the stream lines change their shape forming spirals (cylindrical spiral lines—see in Figure 1.31a).

To describe fluid motion in the TDS control volume, we use the Bernoulli equation valid for an ideal gas

$$v \cdot dp + 0.5d(w^2) = 0. \tag{1.81}$$

Normalizing it via dn (gradient of the normal coordinate) and putting $(dp/dn) = -(1/v) \cdot (w^2/r)$ and $dn = -dr$, we find

$$v \cdot \frac{d}{dn}\left(-\frac{1}{v} \cdot \frac{w^2}{r}\right) + \frac{w}{2} \cdot \frac{dw}{dn} = 0 \tag{1.82}$$

and

$$\frac{dw}{w} = -\frac{dr}{r}, \tag{1.83}$$

where the integration yields

$$w \cdot r = const \quad or \quad w = \frac{const}{r}, \tag{1.84}$$

and r varies along the normal within the range $0 \le r \le R$ ($r = 0$ at the wall and $r = R_0$ at the center). The equation thus derived is valid for an ideal vortex* providing distribution of velocity $w(r)$ along the cylinder radius—see Figure 1.31c. Plots show that fluid velocity is close to zero ($w \approx 0$) in the vicinity of the cylinder centerline, and it is very large at the cylinder wall where $w = \lim(const/r) = \infty$.[†]

Transversely to the stream lines (i.e., along the normal $\vec{n}$) distribution of pressure p is reciprocal to *velocity* distribution: pressure attains minimum at the cylinder wall and maximum—at the cylinder centerline.

The explanation of this phenomenon is based on the following hypothesis: (i) consider gas molecules injected into the cylinder with kinetic energy higher than the mean one; (ii) then, they would start rotating about the cylinder centerline along circular trajectories with larger radius and would naturally approach the cylinder walls; and (iii) particles with lower velocity would tend to occupy trajectories with smaller radius approaching the cylinder centerline.

Thus, radial "stratification" of gas molecules takes place: "*hot*" molecules with high internal energy stand closer to the cylinder wall and "*cold*" molecules with low internal energy

---

* The real flow in the cylinder is not strictly vertical, since particles translate while moving along circular orbits.
† Velocity at the wall is zero in real vertical pipes due to viscosity.

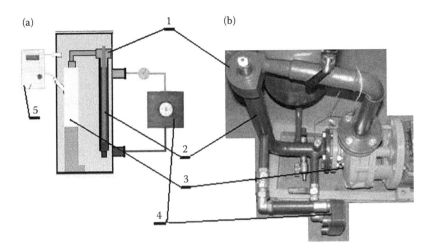

FIGURE 1.32
Vortical heat generator of Potapov: (a) general scheme and (b) physical prototype of a vortical heat generator:
1—tangential vortex generator—intake of the working body (liquid fluid); 2—vortex pipe; 3—pump; 4—heat
exchanger; and 5—ACS with sensors.

stand closer to the cylinder centerline. This "energetic" molecule distribution results in
respective temperature distribution along the cylinder radius as shown in Figure 1.31d.
If the mean temperature at the entrance is T_1, temperature along the cylinder centerline
is $T_{r=0} < T_1$, and temperature at the cylinder wall is $T_{R0} > T_1$. In other words, swirling of
molecules in the chamber divides the working fluid (gas) into two streams—hot (T_{R0}) and
cold ($T_{r=0}$) ones.* The cold stream outflows through a narrow pipe 4 in Figure 1.31d, and
its temperature decreases to $T_{r=0} = -40°C$. The hot current outflows through a regulated
nozzle 5 and its temperature increases to $T_{R0} \approx 110°C$.

The "stratification" of the working body into hot and cold streams can be explained
by the initiation of intermolecular forces and by the dynamics of gas molecule motion.
Although this vortical stream operates under adiabatic conditions, the internal kinetic
energy of the working body close to the cylinder wall increases and the internal potential
energy (pv) decreases, respectively, due to drop in pressure (p). In the vicinity of the cylinder centerline, the internal kinetic energy u decreases due to the increase of the internal
potential energy (p↑), yielding temperature decrease.

Despite that the Ranque vortex tube has relatively low-energy capacity (its unit power
is within the interval 120 ÷ 750 w), and high-pressure source is needed for its operation, it
is used to the air-conditioned compartments of military vehicles where thermal and noise
comfort requirements are moderate.

A modification of the Ranque vortex tube is the "vortex heat generator" of Merkulov
and Potapov[†] shown in Figure 1.32. It is manufactured by OOO"HOTEKA-C" in 7 versions
(ВТГ—5; 7,5; 11; 15; 22; 30; 37; 55 and 75 kW), having high coefficient of conversion of the
mechanical (hydraulic) energy into heat (equal to almost 100%), while the working fluid is
liquid (water) instead of gas.

---

* This phenomenon was discovered by the French physicist Georges J. Ranque in 1928, and Ranque vortex tube
  was patented in 1933 and improved by the German physicist Rudolf Hilsch.
† Professor D.Sc. Yuri Potapov is a member of the Russian Academy of Natural Sciences (RANS), honored
  inventor of the Republic of Moldova.

This special feature totally changes the physical pattern. High-energy molecules enter the peripheral area with high velocity, and a boundary layer with decreased pressure is formed similar to the air vortex tubes. Pressure local value is lower than that of liquid boiling. Hence, intensive boiling and cavitation are initiated in the boundary layer. Although not confirmed, experimental evidence suggests that cold nuclear fusion takes place within the boundary layer and within the Merkulov–Potapov heat generator, converting mass into energy.

If the operation of the generator (the water pump—see position 3 in Figure 1.32) requires for instance 10 kW electric power, the heat exchanger (position 4 in Figure 1.32) releases 15–20 kW of heat.[*]

1.5.6 Control Questions

1. Outline the specific regions of the material state diagram (water state diagram, in particular) considering liquid–vapor transition.
2. How does liquid–vapor transition take place if TDS pressure drops (when does evaporation start and terminate)? What does the energy of phase transition of 1 kg substance depend on (Figure 1.24)?
3. How does phase transition of a liquid proceed at the critical point? Which are the parameters of water's critical point?
4. How does TDS temperature T vary at evaporation? Explain this temperature behavior.
5. Why does TDS temperature increase after completion of the phase transition?
6. Which are the working fluid phases observed during cavitation?
7. When does "boiling" of a liquid start in cavitation areas?
8. What are the effects that occur during cavitation?
9. What are the cavitation-resistant materials for the manufacture of tubes?
10. What are the energy conversions occurring during cavitation? Can they be made use of?
11. How do velocity and pressure vary in a channel with variable geometry—having confusor, diffuser, and straight zones, for instance?
12. What does the Joule–Thompson effect consists in?
13. Which are the general conditions of the initiating the Joule–Thompson effect?
14. How do TDS internal processes run at low temperature ($T < T_{inv}$)?
15. How does internal potential energy vary due to adiabatic gas expansion at low temperature ($T < T_{inv}$)?
16. What happens with internal (kinetic) energy at high temperature ($T > T_{inv}$)?
17. Why do frost or ice cover refrigerator taps or valves mounted after the condensers where liquid flows at high pressure (HP)?
18. What does the Ranque vortex effect consist in?
19. How are velocity and pressure distributed along a Ranque vortex tube?

---

[*] The author's experience has proved that such devices are often offered on the market. Yet, conditions of their manufacture are improvised and the prescribed parameters are not attained.

20. Consider a Ranque vortex tube. How can one explain temperature decrease along the tube centerline and temperature increase at the tube wall?

21. What is the temperature at the exit of a Ranque vortex tube?

22. What is the difference between Ranque vortex tube and Potapov's vortex tube?

23. Which is the real value of the conversion coefficient of Potapov's heat generator?

2

Conversion of Thermal Energy into Mechanical Work (Thermal Engines)

Energy-related (power) technologies may be treated as a combination of engineering-technical methods of energy and work conversion employed to facilitate human life. They are divided into two main groups. The first group comprises technologies of heat conversion into another type of energy (mechanical, electrical, electromagnetic, etc.) while the second one comprises technologies of heat transfer, accumulation, and regeneration. Each thermal technology discussed herein will be illustrated by specific physical schemes and devices. We shall consider them in the following order:

- Technologies of mechanical work performance (so called thermomechanical technologies)
- Technologies of generation of electrical energy (thermoelectric technologies)
- Technologies of heat transformation (regeneration and recuperation)
- Technologies of heat transfer and collection (transfer and accumulation)
- Technologies creating comfortable environment (air conditioning and ventilation)

Thus, we will treat a certain technology as an object of study of respective scientific-applied research fields, on one hand, and we will follow the teaching programs on "Power engineering," "Transport management" and "General mechanical engineering," on the other hand.

2.1 Evolution of Engine Technologies

As is known from physics, energy conversion follows a natural course, that is, energy of motion of macro- and microbodies (popular as mechanical energy) is converted into heat by mechanisms that are studied by tribology (including dry, semi-dry, viscous, or turbulent friction). No opposite transformation is observed in nature. Heat conversion into energy needed for the operation of machines and mechanisms was an impossible task for primitive people as well as for those living in slave-holding* and feudal† societies. Historical records prove that people did not invent a reliable and applicable technology of thermomechanical conversion until the end of the 18th century.

The increase of human population in the Late Middle Ages and the resulting industrial revolution implied the necessity to replace the physical power of people and animals by the power of thermomechanical machines or *thermal engines*. They were harnessed in

---

* Heron's attractive physical models (first century BC) are not considered here.
† Leonardo da Vinci used smoke to set in motion a mechanical gear.

the emerging industry, agriculture, and transport sectors, and their invention affected not only production as a whole, but also product quantity and quality. The development of thermal engines and steam and gas turbines, in particular, yielded significant change of the power technologies, unleashing mass exploitation of electricity in households, industry, and transport.

Initially, the so-called externally fired thermal engines were designed. These were the steam engines of Denis Papin, Thomas Newcomen, and James Watt, the engines of Stirling, the steam turbine of Parsons, and the gas turbine of John Barber. They were known for the heat generation devices (combustion chambers and boilers) that were physically separated from the operating mechanisms (power cylinders and mechano-kinematical chains). Moreover, accessible and cheap heat sources were used—wood biomass and its derivatives (firewood was mostly used till the end of the 19th century), charcoal and coal (their use started at the beginning of the 15th century).

An improvement in thermal engines gathered momentum in the mid-19th century and consisted of efficiency increase and the minimization of engine overall dimensions. The innovative idea was to combine the combustion chamber and the power cylinder into one solid machine unit. This new concept materialized in the "internal combustion engine". Otto and Langen were the first to practically realize the idea in 1858. Later, Atkinson (1882), Rudolf Diesel (1890), and Miller (1947) modified and improved it, while a modern version of the engine is that of Felix Wankel (1951).

Another thermal technology adopting engine partition into functional volumes was designed by Lenoir (1859) and modified by G. Brayton (1872) and Humphrey (in the 1930s). The receiver, compressor, combustion chamber, fuel and flue systems, although mounted in one casing, were physically isolated in separate volumes. The thermal technology was widely applied in gas turbine and aviation turbo-jets, turbo-propeller motors and jets (serial manufacture did not begin until the middle of the 20th century).

The whole 20th century was marked by the improvement of internal combustion engines. Better understanding of the thermal cycles and mechanical construction, and the usage of cheap and easily accessible fuel (oil at the end of the 19th century and natural gas at the end of the 20th century) resulted in the market domination of engines as devices that converted heat into mechanical work. We shall analyze in what follows the basic types of engine technologies that employ the direct thermodynamic cycle as an operational basis for most vehicles. We shall also outline the application of these technologies in industry and in the generation of electricity.

2.2 Engine Technology of Otto–Langen (Carbureted Engine with Internal Combustion)

In contrast to the classical thermomechanical technology of the *Watt engine* (Figure 1.22), where heat generated in an internal combustion device (boiler) was transported to the power cylinder by a heat transferring medium (steam), the *Otto–Langen technology*[*]

[*] N. Otto (1832–91) and O. Langen (1794–1869) patented the carbureted engine in 1861. In 1854–57 Italians Barsanti and Matteucci also proposed an engine operating after the same scheme but their patent was lost and never recognized.

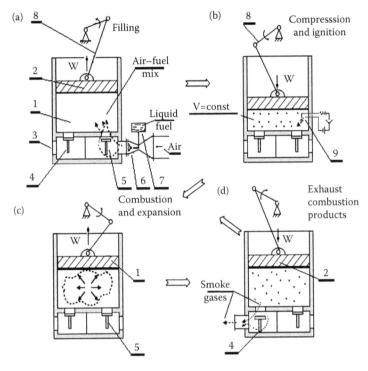

FIGURE 2.1
Carbureted engine operating with Otto–Langen technology: (a) Filling; (b) compression and ignition; (c) combustion and expansion; and (d) exhaust of combustion products. 1—power cylinder/combustion chamber; 2—piston; 3—exhaust passage; 4—exhause valve; 5—intake valve; 6—carburetor; 7—fuel container; 8—rack and pinion mechanism; 9—ignition device.

combined the combustion chamber and the power cylinder into a joint device (see Figure 2.1).

Heat is generated within the power cylinder. Fuel–air mix prepared in a special device (carburetor 6 in Figure 2.1) is fed to the combustion chamber 1. The mix burns in the course of two piston strokes—filling and compression (see Figure 2.1a,b).

A spark device 9 initiates combustion at the end of the second process (stroke), which takes place instantaneously and practically at constant volume (v = const). Then, pressure in the combustion chamber rises almost instantaneously and the expansion process starts from piston top dead center (TDC). Since valves 4 and 5 are closed, expansion is accomplished by pushing the movable wall (piston 2 in Figure 2.1c).

Thus, the engine third stroke, where positive work is done. The piston rises occupying an upper limit position (TDC) and returns back driven by external moments and forces, owing to the engine mechanical components. The thermomechanical system contacts the "cold source" (the atmosphere), while the exhaust valve 4 automatically opens and valve 5 closes. This happens during the fourth stroke of the cycle when piston 2 pushes residue gases which outflow through the exhaust port 3.

The processes are illustrated by the idealized p–v and T–s plots in Figure 2.2. Heat is supplied by the "hot body" during the isochoric process B–1 (corresponding to ignition and combustion of the fuel mix after accomplishing the second stroke), and it is transferred to the "cold body" (the atmosphere) under the other isochoric process A–2 (starting with the initiation of the fourth stroke).

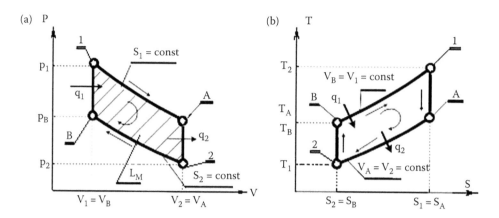

FIGURE 2.2
Idealized illustration of a processes running in a carbureted engine: (a) mechanical diagram; (b) thermal diagram. 1–A—adiabatic expansion (point 1 → point A); A–2—isochoric cooling—clearing of the chamber (point A → point 2); 2–B—adiabatic compression (point 2 → point B); B–1—isochoric heating—ignition and combustion (point B → point 1).

Plots show that the thermal engine operates in a thermodynamic cycle, which consists of 2 isochoric and 2 adiabatic processes corresponding to

- Supply and adiabatic compression of the air–fuel mix from point 2 to point B (Figure 2.2a)
- Air–fuel combustion during the isochoric process from point B to point 1 (release of heat q_1)
- Adiabatic expansion from point 1 to point A and performance of *positive work*
- Isochoric cooling from A to 2 (removal of the combustion products from cylinder)

The cycle is named after one of the designers of the carureted engine—Nikolaus August Otto, and is known as the Otto cycle. Its efficiency is assessed by the coefficient

$$\eta_{Otto} = \frac{L_{II}}{q_1} = \frac{C_v(T_1 - T_A) - C_v(T_B - T_2)}{C_v(T_1 - T_B)} = 1 - \frac{T_A - T_2}{T_1 - T_B}.$$

or

$$\eta = 1 - \frac{1}{\varepsilon^{k-1}}.$$

Here $\varepsilon = v_2/v_1$ is the engine compression ratio.

The above equation shows that efficiency of the carureted engine increases with the increase of ε (v_2/v_1, respectively), and it depends on the engine volume (ε increases with $v_2\Uparrow$).

The Otto–Langen technology was modified by *Rudolph Diesel* (1890) who removed the carburetor. In the diesel engine, fuel is injected at high pressure directly into the combustion chamber at the end of the second stroke. Instead of air–fuel mix, the combustion chamber is initially filled with pure atmospheric air (the first stroke).

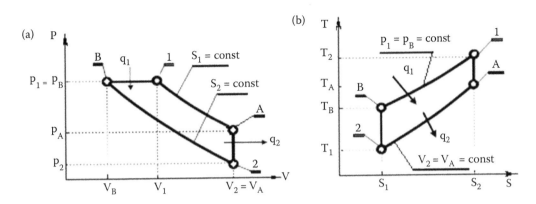

FIGURE 2.3
Diesel engine technology: (a) mechanical diagram; (b) thermal diagram. 1–A—adiabatic expansion (point 1 → point A); A–2—isochoric cooling—cleaning the chamber (point A → point 2); 2–B—adiabatic compression (point 2 → point B); B–1—isobaric heating—ignition and combustion (point B → point 1).

Pure air is compressed from v_2 to v_B along the adiabatic process 2–B (see Figure 2.3a), attaining temperature T_B, which is higher than the fuel ignition temperature. When the piston reaches its lowest position (the lower dead center—LDT), fuel is injected into the chamber at pressure $p_i > p_B$. It mixes with the hot air and burns isobarically (from point B to point 1), while the released heat q_1 causes expansion of the TDS from v_B to v_1. Combustion products (CO_2, CO, NO_2, NO_x, steam, SO_2, etc.) proceed in compressing the piston and expanding the TDS along the adiabatic process s_1 = const (from point 1 to point A—Figure 2.3), doing positive work at the expense of the internal energy (temperature T_1 decreases to T_A). All this happens during the third stroke of the engine operation (from p.1 to p.A).

When the piston occupies the highest position (Figure 2.2d), the exhaust valve 4 opens automatically and spent gases leave the combustion chamber outflowing into the atmosphere at their own pressure p_A and that applied by the piston during the fourth stroke—note that the process is isochoric ($v_2 = v_A$). Thus, similar to the Otto–Langen engine, the diesel engine uses the atmosphere as a cold energy source (TDS cooling takes place via mass exchange between the engine and the surroundings).

Efficiency of that cycle, consisting of two adiabatic (s_1 and s_2), one isobaric ($p_1 = p_B$), and one isochoric ($v_2 = v_A$) processes, is assessed by

$$\eta_{Diesel} = \frac{L_{II}}{q_1} = \frac{p_1(v_1 - v_B) + C_v(T_1 - T_A) - C_v(T_B - T_2)}{C_p(T_1 - T_B)},$$

or

$$\eta_{Diesel} = 1 - \frac{\rho^k - 1}{\varepsilon^{k-1}} * \frac{1}{k(\rho - 1)},$$

where $\rho = v_1/v_B$ is the engine preliminary expansion ratio, $\varepsilon = v_2/v_B$ is the compression ratio while k is the adiabata coefficient.

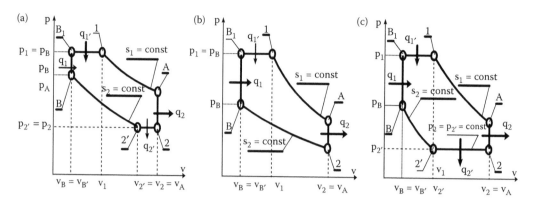

FIGURE 2.4
Improvement of Otto–Langen technology: (a) Atkinson cycle; (b) Sabatier/Trinkler cycle; (c) Miller cycle. 1–A—adiabatic expansion (point 1 → point A); A–2—isochoric cooling—cleaning the chamber (point A → point 2); B–B₁—isochoric heating (point B → point B₁); B–1—isobaric heating—ignition and combustion (point B → point 1); 2–2′—isobaric cooling (point 2 → point 2′); 2′–B; (2–B)—adiabatic compression (point 2′ → point B).

Analyzing efficiency affecting factors, designers found that efficiency essentially depended on the state parameters participating in three ratios, namely

- Engine pressure ratio $\lambda = p_1/p_A$ (Figure 2.3)
- Engine compression ratio $\varepsilon = v_2/v_B$
- Engine preliminary expansion ratio $\rho = v_1/v_B$

The internal combustion engine was compared to the ideal Carnot engine and the problem of increasing its efficiency naturally arose, since efficiency of the early devices barely reached 8%–12%.

Hence, besides R. Diesel (1890), a number of other designers and inventors modified the initial engine technology of Otto–Langen. These were Atkinson (1882), Sabatier–Trinkler (1876–1957), Miller (1947), and many others. Figure 2.4 schematically shows their technological solutions generalizing Diesel's idea.

Considering a TDS doing work at the expense of its internal energy (*u*) and temperature (T), only, the task was to increase the amount of mechanical work done for TDS expansion by "shifting" *point* 1 (the initiation of adiabatic expansion) upward and to the right. A common feature in treating that problem was the attempt to combine the advantages of carburetor and injection combustion in the fuel chamber.

The second qualitative improvement of Otto–Langen engine technology was the decrease of the mechanical work needed for TDS constriction under the Atkinson and Miller cycles (Figure 2.4a,c). This was done by increasing the interaction between the TDS and the "cold source" which yielded the release of additional heat q_2' under isobaric cooling (2–2′) and change of the TDS specific volume from v_2 to v_2'.*

Those improvements resulted in the low cost and high performance of modern cars, Toyota 2004 NHW20 and Lexus HS250 h, for instance.

---

* The technical details of those processes go beyond the bounds of the present book, and they are object of study of specialized literature on internal combustion engines.

The Otto–Langen technology, despite the subsequent improvements of the thermodynamic cycle, has a relatively low efficiency (28%–30%), which is due in particular to the inefficient rack and pinion mechanism, bearings, and cylinder block. Secondly, the thermodynamic process runs under low temperature difference (see Section 1.4.3) which is an obvious physical restriction, and exhaust outflow with relatively high energy potential.

2.3 Engine Technology Developed by G. Brayton

At the time of the Otto–Langen technology, another inventor designed and patented a heat engine in 1872 operating with a different thermodynamic design and with a different structural scheme. His name was George Brayton,[*] an American engineer. The characteristic feature of that engine was that compression and adiabatic expansion took place within two different volumes (i.e., the TDS occupied two different physical spaces—Figure 2.5a,b).

The thermal engine proposed by Brayton has two cylinders. The first one (denoted by 1) serves as a compressor of intake air having parameters p_B and T_B. Air compression takes place along the adiabatic process B–1, while temperature increases from T_B to T_1 and pressure—from p_B to p_1—see Figure 2.5b,c. Air heated to T_1 is supplied to the second cylinder 2 at pressure p_1 through the transfer tube 3 equipped with a switcher. Cylinder 2 operates simultaneously as a combustion chamber and power cylinder. Kerosene (an oil derivative) is used as fuel.

Note that the parameter values of the compressed air within the second cylinder correspond to those in Chapter 1 (v_1, p_1, T_1), and temperature T_1 is higher than the kerosene combustion temperature at the same pressure. Simultaneously, slider 4 of the distributer shifts downward thus blocking the access of hot air to the operating cylinder and redirects the fuel supply to the power cylinder. There, fuel self-ignites burning isobarically, while the TDS expands ($v_1 \Rightarrow v_A$) doing positive work.

After the combustion is completed, the TDS proceeds expanding along the adiabatic process A–2, attaining a specific volume v_2 at the expense of cooling down to T_2. Then, the exhaust valve 5 opens and gases of kerosene combustion gases exit into the atmosphere (the cold source). The efficiency coefficient of that cycle is calculated as

$$\eta = \frac{L_C}{q_1} = \frac{p_1(v_A - v_1) + C_v(T_A - T_2) - p_1(v_2 - v_B) - C_v(T_1 - T_B)}{C_p(T_A - T_1)}.$$

As is easily seen, the Brayton engine has high specific power since it operates along a two-stroke cycle and does the same work as that done by the Otto–Langen engine, but in a shorter time. Tests confirmed its economy and efficiency. Yet, it could not dominate the market because of the strong competition from carbureted and injection engines. However, Brayton engine technology was later employed in the design of two modern thermal engines with continuous operation, namely

- Gas-turbine engines (Figure 2.6)
- Air-jet engines (Figure 2.7)

---

[*] G. Brayton (1830–92).

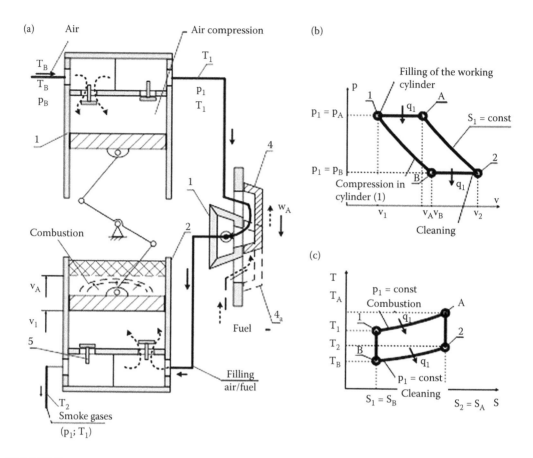

FIGURE 2.5

Two-cylinder heat engine of G. Brayton: (a) arrangement; (b) diagram of the mechanical work; (c) diagram of the thermal work: 1—compressor cylinder; 2—power cylinder/combustion chamber; 3—transfer tube; 4—slider; 5—exhaust valve. 1–A—isobaric heating—ignition and combustion (in power cylinder); A–2—adiabatic expansion (point A → point 2); 2–B—isobaric cooling—exhaust gas removal (from power cylinder); B–1—adiabatic compression (in compression cylinder).

The gas-turbine engine (see Figure 2.6)[*] operates continuously (strokeless), and all cycle strokes proceed successively as in the two-cylinder engine (see Figure 2.5). Yet, the thermodynamic process is organized such as to guarantee maximal useful work done by the output shaft (see Figure 2.6b–e).

The device (Figure 2.6a) has a very compact design and a simplified kinematic scheme. Only *two rotors* with radial grooves (compressor rotor 1 and gas-turbine rotor 3) replace the sliding-rack mechanisms of the two cylinders of the Brayton engine. The combustion chamber 2 is located within the space between the compressor 1 and the gas-turbine 3. Such an arrangement avoids friction losses in rack and slider bearings, piston segments and bolts, simultaneously increasing the engine's angular speed and output power.

The operational order of the thermodynamic processes is kept the same as that in the two-cylinder engine but the arrangement of the technological components is different as

---

[*] The modern gas-turbine engine was designed in 1948 by Joseph Szydlowski (1896–1988), founder of the Turbomeca Company.

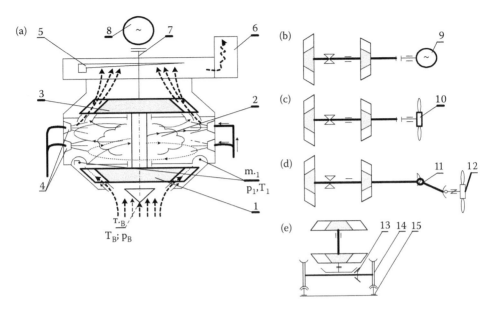

FIGURE 2.6
Gas-turbine engine, coupled with: (a) inside structure; (b) electrical generator; (c) turboprop airplane; (d) ship engine; (e) railway engine. 1—compressor; 2—combustion chamber; 3—turbine; 4—combustion nozzles; 5—gas collector; 6—exaust; 7—output shaft and connector; 8—power application (operating machine); 9—electric generator; 10—airscrew; 11—cardan shaft; 12—propeller; 13—gear; 14—traveling wheels; 15—railway.

shown in Figure 2.6a. The cycle description is similar to that of the two-cylinder engine (see in Figure 2.5b,c)

- Adiabatic compression (B–1) taking place within the compressor 1
- Isobaric expansion (1–A) after fuel supply and ignition in the chamber 2
- Adiabatic expansion (A–2) taking place in the gas-turbine 3 after fuel combustion and finally
- Isobaric cooling (2–B) taking place during the outflow of exhaust into the atmosphere (the cold source) through the gas collector 7 and exhaust 6

Exhaust gases, having released heat to the gas turbine, are released into the atmosphere, while the shaft of the gas turbine does useful work. The gas-turbine engine is used when higher power per unit load is needed, and noise can be limited. Possible applications of the engine are shown in Figure 2.6b–e.

As a whole, the Brayton thermal technology is highly efficient especially when accompanied by co-generation or recuperation of the energy of burnt gases. Quite often, the efficiency coefficient of the installations equals to 80%–85%. To manufacture the technological equipment, however, one needs a structural material that resists high temperature and an aggressive environment. Yet, this is technologically costly and hampers the installation payback.

The second important application of Brayton technology is in *aircraft jet engines* whose schemes and diagrams are shown in Figure 2.7. Its specific feature is that the thermodynamic cycle is organized such that heat is converted with maximal efficiency into kinetic energy of the outflowing gases. Hence, maximal jet draught is attained.

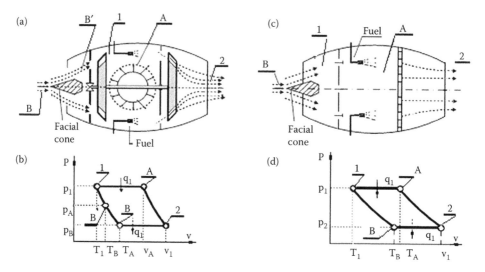

FIGURE 2.7
Air-jet engines operating after Brayton technology: (a) gas-compressor jet engine; (b) mechanical diagram in gas-compressor jet engine; (c) compressorless jet engine; (d) mechanical diagram in compressorless jet engine. 1–A—iochoric heating—ignition and combustion (point 1 → point A) A–2—adiabatic expansion (point A → point 2); 2–B—isochoric cooling—exhaust of combustion products from the chamber (point 2 → point B); B–1—adiabatic compression (point B → point 1); B–B'—adiabatic compression in the defusser.

The following two types of jet engines shown in Figure 2.7a–c are mostly applied:

- An engine with gas compressor
- An engine without compressor

Their common feature is that both engines have adopted the classical Brayton scheme. Isobaric fuel combustion takes place in the chamber (from point 1 to point A)—see Figure 2.7b. A common component is the facial cone, which generates diffuser effect (velocity w↓, and pressure p↑) and compresses the working fluid.

In so far as the facial cone of the gas-compression engine generates auxiliary, preliminary compression of the incoming air (compression from p_B to $p_{B'}$ takes place along the line B–B', and point 1 is with higher pressure), compression along line B–1 is basic and unique in the *compressorless* engine (see Figure 2.7c). Thus, pressure p_1 and temperature T_1 needed for fuel ignition in the chamber are attained. The jet draught R_T at the engine outflow nozzle can be assessed applying the momentum theorem of the theoretical mechanics to the engine working fluid

$$R_T = q_B \cdot w_B - q_2 \cdot w_2 + p_B \cdot A_{In} - p_2 \cdot A_{Out},$$

where

- q_B and q_2, kg/s are mass fluxes at the TDS entrance and exit
- w_B and w_2—speeds of the working fluid at the TDS entrance and exit
- p_B, p_2, A_{In}, A_{Out}—pressure at and areas of the TDS entrance and exit

Maximal draught R_T is attained via an optimization, varying the geometrical characteristics A_{In} and A_{Out}, and the mass flux q_2 of the combustion gases.

Finally, it seems worth noting that a modification of Brayton technology applied in jet engines allows the fuel combustion in the chamber 2 to run under isochoric instead of isobaric conditions.* Then, the engine operates under pulsating (two-state) regime, thus differing from the classical flux jet engine.

Nowadays, due to the improvement of the technologies of steel manufacture and processing, the Brayton direct thermodynamic cycle is applied in large thermoelectric power stations and in small (household) applications. Thus, microturbine co-generators with power ranging from several kW to 100–150 kW are designed, and their capabilities will be discussed in Section 3.2.

2.4 Engine Technology Employing External Heat Sources (External Combustion Engines)

In a historical context, the external combustion engines significantly preceded (by about 400 years) the technologies treated in Sections 2.2 and 2.3. For example, the turbines of Taqi al Din (1551), Giovanni Branca (1629), Gustav de Laval (1893), and Charles Parsons (1894) operated using the energy of superheated steam, but later they gave way to combustion gas turbines.[†]

Yet, they will be further discussed, since they operate after the Rankine cycle and are employed in modern nuclear power stations (equipped with light water reactors), geothermal electric power stations (type "hot rocks"), and solar electric power stations (type "solar concentrators").

The present section treats an engine technology whose efficiency is very close to that of Carnot ideal thermal technology. It is broadly applied in industry and households where significant waste heat is released. Patented by Robert Stirling in 1816, it was the basis of the design of the heat engine prototype (alpha-version) in 1818. The engine was applied to a water pump and used in a sand-pit. It operated following a straight thermodynamic cycle consisting of two isothermal and two isochoric processes (see Figure 2.8e,f).

The original Stirling engine is equipped with two cylinders (Figure 2.8)

- Hot cylinder (HC) 1 called economizer or recuperator
- Cold cylinder 4 (-radiator CR)

Both cylinders are connected to a tube 3 and form the total working surface of a TDS using air as a working fluid.[‡] The standard thermodynamic process of Stirling runs in two phases

- Expansion of the working fluid (from point A to point 2)
- Contraction (from point B to point 1)

Figure 2.8a shows conversion of heat received by the hot body and change of the heat engine kinematics. As shown, the hot cylinder piston 1 occupies TDC, while the cold cylinder piston 4 is close to the right dead center (RtDC). Thus, the thermodynamilac

---

* The technology is known as the Humphrey cycle.
† As far back as 1500 Leonardo da Vinci harnessed smoke gases to a gas turbine to drive a gear. The first actual gas turbine was designed by John Barber in 1791.
‡ Hydrogen (H_2) and helium (He_2) are also used.

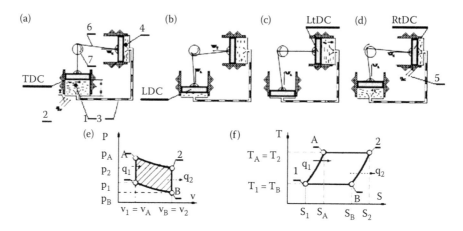

FIGURE 2.8
Stirling engine technology—type Alpha: 1—hot cylinder (HC); 2—hot source; 3—tube; 4—cold CR; 5—cold source; 6—lever mechanism; 7—flywheel. (a) Top dead center of cylinder—TDC$_{HC}$; (b) to LDC$_{HC}$ and LtDC$_{CH}$; (c) lower dead center of the hot cylinder—LDC$_{HC}$; (d) right dead center of the cold cylinder—RtDC$_{CH}$; (e) mechanical diagram; (f) thermal diagram. 1–A—isochoric heating of working fluid; A–2—isothermic espansion; 2–B—isochoric cooling of the working fluid; B–1—isothermoc compression.

system receives heat q_T from the *hot body* through the walls of the recuperator 1 under isochoric conditions ($v_1 = v_A = $ const). Heat consumption is found as

$$q_T' = C_v(T_A - T_1).$$

while TDS pressure rises from p_1 to p_A. Thus, it yields expansion of the working fluid within the connecting pipe and toward the cold cylinder, moving the cold cylinder piston with speed w_2 to the lower dead center (LDC) (Figure 2.8a). Meanwhile TDS temperature tends to $T = T_A$ (the temperature of the external hot body).

Next, isothermal expansion of the working fluid follows ($T_A = T_2 = $ const). It expands from point A to 2 at the expense of a new portion of heat q_T'' received from the hot cylinder during piston motion from top TDC to lower LDC (see Figure 2.8b). Since that expansion proceeds without change in the TDS temperature ($du = 0$ and $\delta q = \delta l_M$), the mechanical work done for TDS expansion from point A to point 2 is equal to the received heat q_T''

$$l_{M_{A-2}} = q_T'' = \int_A^2 pdv = R \cdot T_2 \ \ln\frac{v_2}{v_A} = R \cdot T_2 \ \ln\frac{p_A}{p_2}.$$

When piston 4 of the cold cylinder approaches the left dead center (LtDC—Figure 2.8c) the TDS arrives at point 2, and temporary equilibrium is established in the state diagrams, since the piston speed drops ($w_2 \Rightarrow 0$) and the piston stops for a moment. This is the end of the first stage of the Stirling thermodynamic cycle.

Then, intensive TDS cooling (through the walls of the radiator—the cold cylinder) starts with intensity q_C of the heat flux directed to the cold external source. Isochoric heat exchange ($dv = 0$) takes place in the vicinity of the piston LtDC (position 4 in Figure 2.8c)

$$q_C' = -C_v(T_2 - T_B).$$

The TDS temperature decreases from T_2 to T_B as a result, and no mechanical work is done ($l_{M_{2-B}} = 0$).

When the temperature becomes equal to T_B, that is, to the cold body temperature T_1, the system proceeds contracting isothermally ($T_B = T_1 = const$), and its volume decreases from v_B to v_1.

The work done for TDS contraction is negative

$$l_{MB-1} = q_c'' = -R \cdot T_1 \quad \ln \frac{v_1}{v_B} = -R \cdot T_1 \quad \ln \frac{p_B}{p_1}.$$

TDS pressure rises from p_B to p_1, where the piston of the hot cylinder 1 moves upward, turns the flywheel 7 via lever 6, and the piston of the cold cylinder 4 shifts with increasing velocity ($w_2 \Uparrow$). The TDS constricts isothermally to point 1, and the second phase of the cycle is completed. The process proceeds until a temperature difference between the cold and the warm body exists.

Efficiency of the Stirling engine is given by

$$\eta = \frac{L_C}{q_1} = \frac{l_{MA-2} - l_{MB-1}}{(q_{1-A} + q_{A-2})} = \frac{R \cdot T_2 \ln(v_2/v_A) - R \cdot T_1 \ln(v_1/v_B)}{C_v(T_2 - T_1) + RT_2 \ln(v_2/v_A)},$$

and practically, it does not exceed 12%–15%. This is due to the inefficient convective heat transfer within the boundary layer and to the local effect of viscous loads at the wall during engine operation.

Regarding the counterclockwise Stirling cycle, where the task is to change the temperature of a hot body at the expense of external work done, energy efficiency is assessed via the so called coefficient of performance (COP)

$$COP = \frac{q_2}{L_C} = \frac{q_{2-B}' + q_{B-1}''}{L_C},$$

When Stirling technology is used to change the TDS parameters as a result of the thermopump effect, efficiency is assessed using the conversion coefficient defined as

$$\eta_{HP} = \frac{q_1}{L_C} = \frac{q_{1-A}' + q_{A-2}''}{L_C}.$$

Stirling thermal engines are recommended to participate in schemes of waste heat recuperation adopted by a wide variety of industrial machinery and equipment condensers of thermoelectric power stations, nuclear power stations or air conditioners, electrical transformers of draught and regional substations, walls of combustion chambers, boilers flue, etc.

Besides the Stirling engine, there are other types of engines of that sort (alpha-, beta-, and gamma Stirling modifications, rotary Stirling engine, Fluidyne engine, Rigban engine patented in 1907, etc.). In 1951, the Philips Company was the first to manufacture a prototype of an encapsulated Stirling engine. Yet, large-scale engine manufacture was not realized due to the high cost.[*] The first Stirling refrigerator using a counterclockwise cycle was manufactured in Japan in 2004 by Twinbird Corp.

---

[*] A series of 150 pieces were manufactured.

Due to the relatively simple mechanical equipment, the Stirling technology awaits new renaissance in the manufacture of combined steam power generators with solar concentrators.

2.5 Mechanical–Thermal Technologies (Gas and Vapor Compression)

So far, we showed that mechanical energy naturally converts into heat. For instance, applied mechanics proves that significant heat quantity is spontaneously released during the change of the parameters of moving mechanical systems and due to the action of resulting friction forces and moments. This holds true for mechanical gears (gear reducers or multiplicators, belts, friction devices, chains, etc.) where heating is inevitable, unacceptable, and even damaging. In other cases discussed in the previous sections (those involving van der Waals forces, for instance) even the mechanical motion of atoms and molecules is converted into heat as is in cavitation, throttling, or potential flow vortexing. There, heat release is spontaneous and uncontrollable.

However, directed conversion of "external" mechanical energy (work) into heat can be performed in a TDS. This would result in TDS heating along a counterclockwise thermodynamic cycle. Compression is an example of such a process, taking place in a number of energy-converting (power) machines, vapor compressors, air refrigerators and heatpumps, carburetor and injection engines with internal combustion, gas turbines, jets, etc.

Gas and vapor compression is essentially important for the total efficiency of thermodynamic devices. The treatment and clarification of that process is an important issue of specialists trained in design, exploitation, and service of power machines and installations. However, we shall discuss the operation of the counterclockwise cycle of gas and vapor compression and conversion of the mechanical work into heat, without delving into detail. At the same time, we shall present a short survey and classification of compressor types.

2.5.1 Compression Cycle

As noted, compression of gases and vapors runs actually along a counterclockwise thermodynamic cycle where an "external" mechanical work is done on a TDS. It consists of four phases (see Figure 2.9b)

- Compression (point 1–point A)
- Extruction (point A–point 2)
- Expansion (point 2–point B)
- Suction (point B–point 1)

Efficiency is given by the traditional ratio

$$\beta_{Compr} = \frac{\text{useful effect}}{\text{inserted energy}} = \frac{\text{gas internal energy}}{\text{external work}},$$

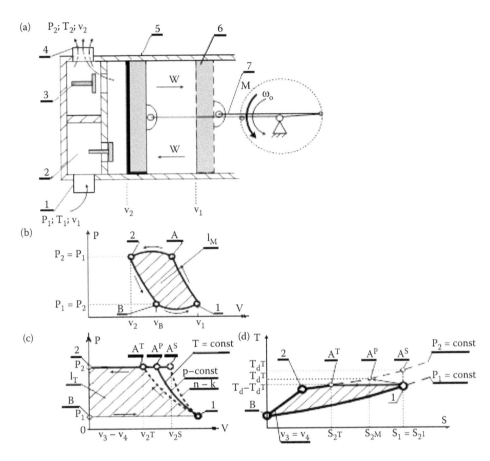

FIGURE 2.9
One-stage compression: (a) scheme of a single-cylinder piston compressor: 1—intake port; 2—intake valve; 3—exhaust valve; 4—exhaust port; 5—piston at left dead center; 6—piston at RtDC; 7—slider-crank mechanisms; (b) indicative p–v plot; (c) idealized p–v plot; (d) idealized T–s plot. 1–A^S—adiabatic compression; 1–A^P—isobaric compression; 1–A^T—isotermic compression; A–2—isobaric compression; 2–B—isochoric pressure normalization; B–1—isobaric expression.

and it is interpreted assuming that (Figure 2.9b,c)

1. All processes are reversible
2. Compressor parasitic volumes are neglected (i.e., $v_3 = v_4 = 0$)
3. Phase 2 (extrusion A–2) and phase 4 (suction B–1) are executed under isobaric conditions ($p_A = p_2$ and $p_B = p_1$)

The idealized counterclockwise thermodynamic cycle of gas and vapor compression is illustrated in Figure 2.9c,d.

The first diagram shows that the mechanical work needed for the total work cycle (the hatched area fixed by points 1, $A^{T;n;s}$, 2, and B) depends on the process start-compression. A minimal amount of mechanical work is needed if compression takes place under isothermal conditions ($dT = 0$ and $T = const$), and maximal amount of mechanical work is needed for adiabatic compression ($ds = 0$).

For real compression of gases and vapors (under polytropic conditions L_C^n from point 1 to point A^n), the amount of mechanical work needed to cover a single cycle is within the following specific range:

$$L_C^T < L_C^n < L_C^k.$$

The plot in Figure 2.9d is an idealized T–s diagram for the expected "useful" effect—the thermal work done throughout the entire cycle (the hatched area is proportional to the heat generated within the cycle).

Using this area and the end points of compression–point A (A^T, A^n, and A^s) one can predict the efficiency of "mechanical work-heat" conversion.

Under isothermal compression, the total mechanical work converts into heat but is not sufficient to heat the working fluid, since $u_{1-2} = 0$, which maintains its temperature. Then, the thermal efficiency of the cycle is $\eta_{Compr} = 0\%$, since the system is open and replacement of the working fluid loss takes place.

As for adiabatic compression ($1–A^s$) the total mechanical work is done for the increase of gas internal energy, pressure, and temperature. Hence, the working fluid (gas or vapor) is most effectively heated within an open TDS under adiabatic conditions (since $u_2 = u_1 + L_C$). Besides, the working fluid accumulates the total input work, and its temperature increases. The newly attained temperature T_2 is calculated as $T_2 = T_1 + L_C/C_v$.

Practically, compression yields increase of the working fluid temperature and pressure (i.e., body potential energy is made use of). This is achieved in hydraulic and air transfer of mass or energy. Hence, we shall consider different types of mechanical equipment for gas and vapor compression.

2.5.2 Classification of Compressors

Compressors are systems made up of an external motor (electric motor, heat, gravitational, or hydraulic engine) and an executing mechanism (slide-crank, disk, rotor or coupled rotating shafts, spiral disks, etc.). Their task is to pressurize the working fluid (air, refrigerator vapors, etc.). Compressors are classified according to two criteria

- Principle of operation
- Exhaust pressure

Regarding the first criterion, there are volume, centrifugal, and ejection compressors.

Volume compressors pressurize the working fluid changing its initial volume. They are piston type (Figure 2.9a), eccentric type (Figure 2.10a), screw type (Figure 2.10b), and scroll type* (Figure 2.10c). All of them have hermetically closed control surfaces with one or more movable segments (movable wall, piston, and surface) driven by an external motor.

As seen in Figure 2.9a, the piston acts as a component of a kinematic pair (position 5). It performs translation as a link of a slider-crank mechanism. In eccentric compressors, the working fluid 4 in Figure 2.10 is forced to move in a channel with decreasing volume bound by an eccentrically located rotor (rotating wheel), fixed within the compressor casing, and the casing inner surface (position 1). Constriction of the TDS control space is attained by rotation of compressor screw and spiral shafts shown in Figure 2.10b,c.

---

* Invented in in 1905 by Louis Creux, but industrially manufactured after 1980 by Volkswagen.

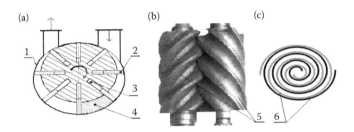

FIGURE 2.10
Volume compressors: (a) eccentric compressor; (b) Lisholm screw compressor; (c) scroll compressor: 1—fixed cylinder (casing); 2—sliding seals; 3—channel; 4—volume compressed by the working fluids; 5—rotating screw couple; 6—movable and immovable spiral screw.

Volume compressors convert the kinetic energy of their movable segments into mechanical work needed for segment shift thus decreasing the TDS control volume (i.e., performing TDS contraction). This *pressurizes* the working fluid, and the TDS internal energy increases. Volume compressors operate within wide pressure and capacity ranges. The typical range of the limit pressure is 0.1–3.0 MPa, and the capacity of rotary compressors is about 2000–4000 m^3/h.

Centrifugal compressors pressurize the TDS working fluid due to the normal d'Alembert forces. The latter occur during fluid motion in the curvilinear rotating channels of the compressor rotor. The centrifugal compressors are classified with respect to the geometry of their rotors. There are compressors whose blades are straight or bended (forward or backward—see Figure 2.11).

According to the second criterion, compressors are divided into three groups

- Low pressure compressors (LP)—for $p_2 \leq 0.11$ Mpa (ventilators with medium pressure (MP) and high pressure (HP) are also part of that group)
- MP compressors for $p_2 \leq 1.2$ Mpa
- HP compressors for $p_2 \leq 1.2$ Mpa

One-stage centrifugal compressors operate at maximum pressure ranging up to about $0.8 \div 1.0$ MPa, and multistage compressors—at pressure of up to 12.0 MPa. Compressors

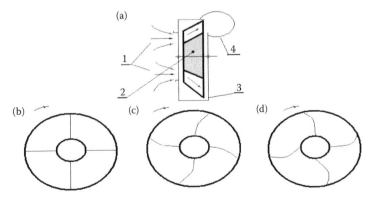

FIGURE 2.11
Centrifugal compressors: (a) schematic section of the working rotor (impeller) of a one-stage centrifugal compressor; (b) impeller with straight blades; (c) impeller with forward slanting blades; (d) impeller with backward slanting blades. 1—intake confusor; 2—impeller; 3—casing; 4—exhaust diffuser.

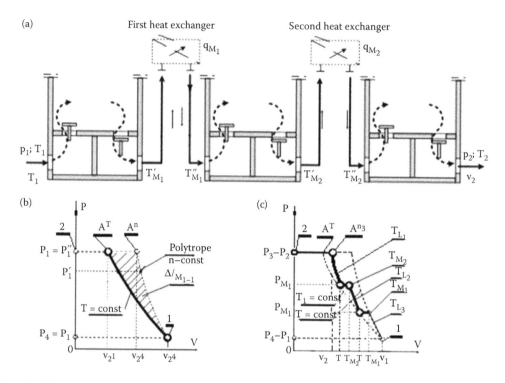

FIGURE 2.12
Multistage compressor with intermediate cooling: (a) arrangement of equipment for multistage compression; (b) the unnecessary work done by the external engine increases; (c) the one-stage polytrope "n" is replaced by 3 polytropes (n_1, n_2, and n_3) to minimize the external work.

operating at medium and high pressure are designed using two- and multistage schemes. The problem that arises is illustrated in Figure 2.12b—increase of the degree of pressure rise $p_2/p_1\uparrow$ yields significant increase of the energy used by the drive engine.

2.5.3 Multistage Serial Compression

Consider a working fluid (air) at initial state 1, with initial temperature (T_1), pressure (p_1) and specific ratio "volume/density" (v_1/p_1), having undergone one-step compression to p_2. Compression has taken place along the polytrope $1 - A^n$.

Figure 2.12b shows that the external mechanical work $l_{M_{1-2^n}}$ done will exceed by $\Delta l_{M_{1-2}}$ the work done for isothermal compression if compression has taken place along the line $1 - A^T$. It is also seen in the figure that the difference $\Delta l_{M_{1-2}}$ in favor of the polytropic process increases with pressure rise, if compression to p_2'' instead of p_2' has been performed (compare the hatched areas $1 - A^{n'} - A^{T'}$ and $1 - A^{n''} - A^{T''}$). To attain output pressure p_2'', a significant increase in primary energy is needed for the compressor operation.

When the task is to reduce the consumption of primary energy at a higher degree of pressure rise ($\eta = p_1/p_2$), multistage serial connection of compressors is performed as shown in Figure 2.12a). Thus, economy of external energy is realized—see Figure 2.12b). It is due to the *intercooling heat exchangers* that operate after each stage. The effect of their use is illustrated by the working diagram plotted in p–v coordinates.

At the first stage, pressure p_1 rises to p_{M_1} (Figure 2.12c). The process runs along the polytrope "n_1" from point 1 to point M_1. At the end of the first compression stage, the working fluid with parameters corresponding to point M_1 (p_{M_1}, v_{M_1}, and T_{M_1}), leaves the first cylinder and enters the first heat exchanger where its temperature decreases to T_{M_2}

$$T_{M_2} = T_{M_1} - \frac{q_{M_1}}{C_v},$$

which is close to T_1.

Compression proceeds within the second and third cylinders along the polytropes "n_2" and "n_3," respectively (Figure 2.12c). To maintain compression close to the isotherm T_1, the working fluid cools down again after the second compression stage (after the second cylinder—Figure 2.12a). The plot "p–v" of multistage compression gains a "step-wise shape" with T_1 as its asymptote. The plot would coincide with T_1 for an infinite number of intermediate cooling periods. Yet, this extrapolation obviously is not practical, since the whole installation would become very complex and costly.

Generally, the number of compression stages z is found via $\lambda_i = \sqrt[z]{p_2/p_1}$, and the necessary compression stages are found as $\Rightarrow z = \ln(p_2/p_1)/\ln\lambda$. Here $\lambda_i = \lambda = \text{const}$ is a coefficient of pressure rise between two compression stages. Pressure gained at a corresponding stage is $p_{M_i} = \lambda_i \cdot p_1$. Pressure gained after the first stage is $p_{M_1} = \lambda_1 \cdot p_1$, that gained after the second stage is $p_{M_2} = \lambda_2 \cdot p_{M_1}$, etc.

Heat released in the intermediate heat exchangers $q_{M_i} = C_v(T_{M_i} - T_{M_{i+1}})$ has a significant energy potential, and it should be recuperated in an appropriate system (see Chapter 3).

Note that engine technologies played an important role in civilization. Moreover, they are expected to occupy a significant position in the further development of transport, industry, and popular culture, since they continuously advance with the evolution of material science and fuel chemistry. In this respect, the design of new highly efficient thermomechanical engines is forthcoming. They will be manufactured from nanomaterials with increased thermal resistance and strength and will operate using environmentally friendly decarbonized and denitrogenized fuel.

2.6 Control Questions

1. Outline the features of thermomechanical technology. What is the direction followed by a thermodynamic process, which runs in such a power machine?

2. Can the efficient coefficient of a thermomechanical technology be equal to or higher than 1? How can one find the maximal efficiency coefficient?

3. What is the difference between a thermomechanical technology using internal combustion and that using external combustion? Give an example of their application.

4. What is the difference between the engine technology of Otto–Langen and that of Brayton?

5. How did Rudolph Diesel modify the engine technology of Otto–Langen? Which are the advantages of that modification?

6. What is the difference between the engine technology of Otto–Langen and that of Atkinson? What did Atkinson aim for with his modifications?

7. Which are the advantages and disadvantages of the engine technology of Brayton? Why didn't the two-stroke engine prevail on the market? What was its fuel?

8. What is the difference between the gas-turbine heat engine and the turbo-jet heat engine? What is the similarity between them?

9. Consider a steam turbine and a gas turbine. Which are the technologies (with internal or external combustion) that they employ?

10. How many types of Stirling engines do you know? What is the difference between each specific type? Which are their advantages as compared to other engines?

11. Why wasn't the Stirling engine widely applied? What is the area of its application?

12. Which thermodynamic cycle is used to describe compression—the direct cycle or the counterclockwise cycle? When does the most effective "mechanical energy-heat" conversion take place?

13. Why is isothermal compression preferred in compressors of industrial installations?

14. How can one technically attain "almost" isothermal compression?

15. How can one utilize the "waste" heat released in intermediate heat exchangers during multistage compression?

16. How many compressors do you know with regard to their principle of operation?

17. How many compressors do you know with regard to output pressure?

3

Thermoelectric and Co-Generation Technologies

This chapter discusses some energy-conversion technologies where heat is directly converted into electricity or is a previously created product by energy conversions. We shall analyze the following technologies:

- Conversion of solar radiation into electricity (so-called "internal ionization")
- Seebeck technology applied in thermos-couples (so-called "thermoelectric current")
- Schonbein/Grove technology with oxidation control
- Rankine cycle technology
- Brayton cycle technology

Historically, static electricity[*] was known since medieval times and much earlier (Thales described some static electricity effects in the seventh century BC). By mid-16th century however no one in Europe distinguished electricity from magnetism. This was done by Gerolamo Cardano (1501–1576) in his treatise, thus stimulating the interest of other European researchers in the phenomenon and laying the foundations of electric generator design.[†] Following the invention activity in the field, we may outline two methods of heat-electricity conversion:

- *Direct method* (DTEG-direct thermoelectric generation)
- *Indirect method* (InDTEG)

Technologies using internal ionization and barrier semiconductor layers resulted from the first method, including photovoltaic[‡] cells which operate using radiation of hot bodies (the Sun especially), as well as thermoelectric converters exploiting the Seebeck effect for direct conversion of heat into electricity. Other technologies of the same type are those applied in fuel cells (FCs),[§] where electricity is generated during a controlled process of oxidation of gaseous hydrocarbon fuels, such as natural gas, syngas, hydrogen, etc.

Currently, the second group of technologies (InDTEG) dominates in all industrial installations for central generation of electricity in electric power stations. Their characteristic feature is the performance of a cascade of intermediate energy conversions (see Section 1.1.6). For instance, burning (*uncontrollable oxidation*) generates heat as *a first step*. Then the

---

[*] The term was introduced in 1600 by W. Gilbert (1544–1603).

[†] The first electric generator was designed by A. Volta (1745–1827) in 1800, and it converted chemical work into electrical energy.

[‡] The photovoltaic cell operates according to the principle of internal ionization discovered by Becquerel (senior) in 1839. The first photovoltaic cells were manufactured in Bell Labs in 1954.

[§] Oxidation in a catalytic environment was first described by the German physicist K. Schonbein (1938) but the first industrially applicable and commercialized technology for the generation of electricity was designed in 1842 by W. Grove (1811–1918).

heat is converted into mechanical energy of the kinematic chains of internal or external heat engines, which are aggregated to the electric generators. *Finally,* the generators convert the available kinetic energy of the rotating or translating electric induction devices[*] into electricity.

Some modifications of InDTEG[†] are the so-called nuclear and solar technologies which convert *radiation* (resulting from the nuclear decay of processed uranium or coming from the Sun) into heat carried by superheated steam, and the heat is *finally* converted into electricity.

3.1 Direct Thermoelectric Technologies

3.1.1 Internal Ionization of Bodies and Photovoltaic Cells

To clarify the mechanisms of internal ionization of bodies one must remember the principle of operation of Einstein atomic oscillators in a crystal lattices (see in Figure 1.2), considering the zone theory of atom and molecule structure. Accordingly, the electrons of each atom occupy different orbits (energy orbitals). Their passage from one to another orbital would be possible if only the electron receives or releases discrete portions of energy called quanta. The number of electrons in an orbital around the atomic nucleus is also limited (not more than two electrons in the first orbital, not more that eight electrons in the second one, and then not more than 18, 32, 50, etc. electrons). The last and remotest energy orbital is called valence band, and it is not filled with electrons.

Electrons may leave their orbits, break free from their atoms and exist independently (move within the interatomic space) forming an electron gas. To start a free motion, *valence band electrons* should overcome the energy barrier of the "banned zone" (the last energy hole between the quantum orbitals, which the electrons should "jump over" prior to their complete emancipation from the atoms). The process of the electron's "jumpover" the banned zone and shift to the free zone is known as "lattice internal ionization," since atomic nuclei are converted into positive ions (anions). A "hole or vacancy" is formed as a replacement of an electron (it behaves as a positive charge—positron).

Solids, being medium where energy conversions take place, can be classified into three groups, depending on the "width" of the banned zone and on the possibilities of "jump over"

- Insulators (additional energy exceeding 5 eV are needed to overcome the energy gap of the banned zone)
- Semiconductors (the banned zone is narrow $E_g < 5$ eV)
- Conductors (the electrons of the valence band pass freely to the free area, since no banned zone exists in the valence band and in the free area—both energy orbitals merge)

---

[*] The first electrostatic generator was designe by Otto von Guericke in 1650. The first patented one using electromagnetic induction for its operation was the single-pole disk generator of Michael Faraday (1831), while his industrial prototype was manufactured in 1832 by H. Picsius (1808–1835). However, those devices were not commercially realized.

[†] Even technologies used in wind power stations or in hydroelectric power stations are InDTEG, since the initial energy is solar radiation which participates in the gigantic energy-conversion mechanisms of the Earth atmosphere (described in Chapter 3).

TABLE 3.1

Width of Banned Zone for Different Semiconductors in eV

Substance	Si	Ga	Sn	GaAs	InSb	HgSe	InP	GaP	ZnSe
Width of the banned zone, eV	1.107	0.67	0.08	1.35	0.165	0.3	1.27	2.24	2.58

Semiconductors are the most interesting for devices converting energy. The energy barrier of their banned zone is easily overcome using "natural" energy carriers, that is, solar radiation photons or those emitted by bodies heated in a combustion chamber (CC). Their energy is proportional to the fourth degree of temperature—T^4 (the law of Stefan–Boltzmann). This is clear from the comparison between Table 3.1 and Figure 3.1.

The energy-converting medium specified in Table 3.1 belongs to the group of semiconductors with narrow "banned" zones. It is seen that they have very low energy barriers of the banned zone varying in the range 0.08–2.58 eV. On the other hand, Figure 3.1 illustrates the energy of the "solar" photons and their ability to displace electrons from the valence band depending on the wave number of the carrying orbitals.

It is seen in the figure that orbitals with wavelength within ranges $350 < \lambda_{Ph} < 760$ nm possess the largest number of photons, and hence—the greatest energy charge. They form the visible part of the solar spectrum. Photon energy charge is within limits $1.7 < e_{Ph} < 3.1$ eV. The energy is sufficient (see Table 3.1) for valence electrons move over the banned zone and leave as a cloud of free electrons. At the same time, the atomic lattice of the body is ionized within the area which the photon stream falls on (part of the body external surface).

Although the phenomenon was already known in the first half of the 19th century, inventors did not manage to design a direct converter of solar energy into electricity.

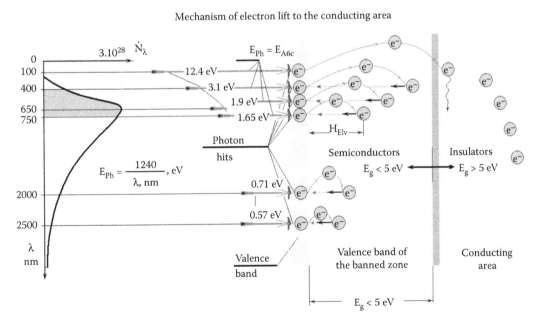

FIGURE 3.1
Examples of overcoming the banned zone.

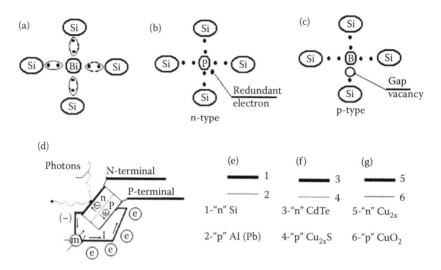

FIGURE 3.2

Photovoltaic structures: (a) silicon lattice (IVth group); (b) Si "n-layer"; (c) Si "p-layer"; (d) cross section of a photovoltaic "n–p" cell; (e) Fuller cell (Si–Si); (f) cell with different bases (Cu_2S/CdS); (g) Shottky cell (1930)—thin barrier ("n" Cu_{2x}/"p"CuO or "n"Si/"p" Si) with a thin metal layer.

The first photosensitive element was developed 130 years ago, and it became clear that the direction of research was correct. Yet, the research "break through" was made during the 1950s using the so-called "n–p"—silicon semiconductors which were the first devices to successfully convert solar energy into electricity.

What is the construction of a silicon semiconductor and how does it operate under solar photon radiation?

Figure 3.2a–c presents as an illustration three structures of atomic lattices built on a silicon basis. The difference between them is evident when comparing the first structure to the second and the third ones. It is seen that when an atom from the natural configuration of the silicon lattice is replaced by an atom with higher valence (phosphorus from group V, for instance) an extra valence is left within the lattice (an extra atom). A similar result occurs when replacing an atom from a lower atomic group (as in Figure 3.2c—Boron from group III). Then, an atom abandons the crystal structure leaving a so-called "hole or vacancy." It is assumed that bodies with such an atomic structure should be called bodies with "n" and "p" electric *conductivity*, due to electron excess and shortage.

Two very thin "n" and "p" layers are arranged along the direction of supply of a modern photovoltaic cell* (see Figure 3.2d) with solar energy (photons from the visible and infrared regions of the spectrum). So far, a number of "n–p" combinations have

---

* The cheapest cells are those of "*amorphous* Si." Yet, they operate with the least efficiency (about 8%) under direct radiation. At the same time, they display affordable efficiency when operating in countries with moderate climate due to the specific climatic conditions (availability of large portion of diffusive radiation and variable temperature). More expensive cells are manufactured from *monocrystal* silicon (having 16% efficiency) and thin-layer *polycrystal* silicon (with 14%–15% efficiency). Their higher price makes them inappropriate for mass usage. The so-called "multilayer" cells display significantly larger efficiency (~43%) in laboratory conditions.

been designed using various chemical elements—some of them are shown in Figure 3.2e–g.[*]

How do the two "n" and "p" layers operate jointly under direct solar radiation? Solar photons penetrate both semiconductor layers "bombarding" the electrons of their valence bands.

If the photon energy absorbed by the electrons is larger than the energy threshold of the banned zone, they leave the atoms of the layer and start moving in the free conducting zone (if the energy is not sufficient to overcome the energy barrier of the banned zone the electrons stay within the orbital surrounding the atomic nucleus. They send back the received photons in an arbitrary direction and give birth to phonons through which energy dissipates as heat). Some electrons in the "p layer" move to the "n layer," creating significant electron deficit in the "p" layer and significantly increasing the positive potential of the "p layer." At the same time, electron excess occurs in the "n layer" and its negative electric potential increases (the crystal lattices of both layers are internally ionized).

If both layers are connected to instruments via closed circuit (see in Figure 3.2d), electromotive force (emf) arises under the potential difference and electric current flows. The electrons leave the "n layer" and return to the "p layer" through the closed external circuit where they recombine with lattice anions. The process does not damp since the solar photons leave vacancies in other lattice atomic structures. Since the voltage at the cell exit under normal conditions ($T_e = 293$ K and $I_{solar} = 1000$ W/m^2) is very low (1.36 V), cells are assembled in solar panels consisting of 25–36 individual cells with total power of up to 230 W (see Figure 3.3b). On the other hand, solar panels are assembled in series with output voltage of 400 V called *"strings."* The electric energy generated is usable after conversion (DC→AC) of its parameters by means of electronic devices called inverters. The converted parameters should correspond to those of the final users.

The efficiency of the existing laboratory prototypes of multilayers photovoltaic cells is good—37%–43%, but just 8%–15% of the incident solar energy is converted into electricity by means of the commercial *PV cells*. The rest energy spontaneously dissipates in the surroundings as heat.

Solar internal ionization is a phenomenon where two different (co-generation) energy forms are obtained—electrical and thermal (see Dimitrov, 2015). Recently, a tendency to integrate the solar cells with thermal panels ("n–p" plus thermoabsorption hydraulic system), called hybrid panels has occurred. This is expected to yield better operational

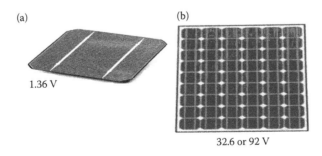

(a)

1.36 V

(b)

32.6 or 92 V

FIGURE 3.3
Photovoltaic structures: (a) cell with output voltage 1.36 V; (b) panel—32.6/92.0 V.

---

[*] The new photovoltaic cells are based on plastics, polymers, and bio ferments. These are cells manufactured from conducting plastics designed by A. Heeger (Nobel Prize winner 2000), cells manufactured from conducting polymers Olotrau of NSF, and chlorophyll cells of Gratzel cells.

conditions both for the *electric/thermal* system with total conversion efficiency in the range 71%–82%, especially in winter.

Regarding the geography of Southern Germany, electricity of 8.0 kWh per day can be produced by a photovoltaic cell with 2 m² surface (cell productivity for Southern Europe is about 10.0–12.0 kWh per day). If such a cell is integrated with the building envelope, its energy efficiency will rise due to co-generation.

Prices of photovoltaic systems were about $6000/kW in 2005, while now they are less than $2500/kW 2500. The prime cost of 1 kWh generated by a photovoltaic system was $0.76/kW in 2005. Yet, it is now significantly lower, and it is expected to drop to $0.17/kWh in 2030. This will make the photovoltaic technologies fully competitive with thermoelectric technologies, especially in small (household) applications. Until then however, they should be treated with special care by officials due to their friendliness to environment and compatibility with the central systems for energy generation and distribution.

3.1.2 Thermoelectric Technology of Seebeck

The German physicist Thomas Seebeck (1770–1831) found in 1821 that when two thermo/electro conducting materials with different conductivities were formed together, a voltage difference (ΔV) occurred between their free ends—Figure 3.4a (at present, the combination is known as a thermocouple). At the same time, Seebeck established that the value of the electro-potential difference (ΔV) depends on the temperature of the environment where the thermocouple is located.

In 1823 Seebeck showed that if two thermocouples were serially connected, electric current ran through their wires upon immersion in warm (T_1) and cold (T_2) physical medium (energy sources), respectively (Figure 3.4b).

He found that emf occurring in the circuit was directly proportional to the temperature difference $\Delta T = (T_1 - T_2)$, i.e.,

$$emf \approx \Delta T,$$

that is, if $\Delta T \uparrow$, then emf $\uparrow$, and vice versa.

The relation between temperature difference dT and the electric potential difference dV in deferential form reads as follows:

$$dV = S_{AB} \cdot dT,$$

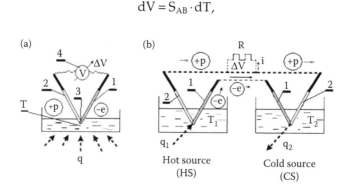

FIGURE 3.4
Seebeck effect: (a) demonstration; (b) thermoelectric circuit: 1—copper rod (n-type); 2—constantan rod (p-type); 3—contact weld; 4—galvanometer.

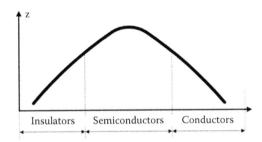

FIGURE 3.5
Optimization of the complex $Z = S_{AB}/(\lambda\Omega)^{0.5}$ for different materials.

The above relationship is known as the Seebeck equation.

It is found that the proportionality coefficient (S_{AB}), named after Seebeck, depends on the thermocouple materials. It has been proved that the value of the Seebeck coefficient (S_{AB}) is proportional to the coefficients of thermal and electrical conductivities (λ and Ω)

$$S_{AB} \approx \sqrt{\lambda \cdot \Omega}.$$

Although that the link Z ($Z = S_{AB}/(\lambda\Omega)^{0.5}$) has a clear mathematical structure, it is a complex physical quantity depending on the capability of metal crystal lattices to react to external thermal impacts—this is the lattice capability to interact with infrared photons whose carrying wavelength exceeds 860 nm. The results in Figure 3.5 for the values of the functional $Z = S_{AB}/(\lambda\Omega)^{0.5}$ show that semiconductors are materials most appropriate to realize Seebeck technology.

After Thomas Seebeck, Jean Peltier in 1834 and William Thomson (Lord Kelvin) in 1854 obtained similar results. They studied different heat convertors of the "n–p," type, Bi_2Te_3 or Cu-GeSi for instance, thus finding inverse thermoelectric effects.[*] Initially, the thermoelectric effect of Thomas Seebeck was explained by the different conductivity of the couple materials. Later, it became clear that this was a new phenomenon called "internal ionization," while can be explained by the zone theory.

Now, 200 years later, it is clear that the Seebeck effect is due to the thermocouple specific atomic structure and its capability to undergo different internal ionization under the impact of photons radiated by heat sources. The processes observed and described by Seebeck can be explained as follows (Figure 3.4).

The copper rod being an "n"—type conductor conducts heat according to Kaganov's mechanism.[†] Copper is a metal belonging to group II of Mendeleev's table. It has two electrons in its valence band, and its banned zone merges into a free zone. Electrons are released from the atomic nuclei of the crystal lattice spontaneously and under low temperature. When a contact with the hot source is realized, the atoms of the contact area get

---

[*] Thermoelectric process is convertible. If two "n–p" contact couples are serially connected and switched to an external electric source, the electric current flowing is converted into heat (the cold body cools down—Peltier effect, and the hot body warms up—Thompson effect).

[†] Kaganov's model called "two step model" is designed on the basis of e phenomenological mechanisms of interaction between phonons and electrons. It consists of two phases: heat transfer is realized via the directed motion of electrons of the free zone (without emf) during the first phase; the second phase consists of transfer of the kinetic energy of the moving electrons to the lattice atoms which start oscillating (generate phonons) with frequency v depending on the temperature of their equilibrium state t_l (static potential).

excited and internally "ionized" under the effect of long wavelength photons. Besides, the atoms of the valence band pass over the banned zone (it is very narrow for Cu) and form a gas of free electrons belonging to the crystal lattice of the contact area between the thermocouple and the hot source. The quantity and concentration of the free electrons depends on the hot source temperature T_1. Simultaneously, the second contact thermocouple at the other end of the circuit is immersed in the cold source which temperature is T_2. Due to the low temperature, electron concentration of the free electron gas is significantly lower, since the process of internal ionization is less intensive.

Since both ends of the copper rods ("n" type) are connected by a cable, the electrons of the "hot" end flow to the "cold" end (see in Figure 3.4b). If there is no second cable, equilibrium will take place after the equalization of the electron concentration at the two rod ends, and the electric current will be blocked.

Regardless of the transfer of "n" electric charges through the copper cable, similar processes* run in the other two rods manufactured from semiconductors with "p" conductivity (see Figure 3.4b). Yet, holes (vacancies) are emitted to the areas of contact with the heat sources. The difference is that more holes will concentrate around the cold body than around the hot body. Those holes have diffused through the weld and have been directed to the "n" zone.

If only rods with "p" conductivity are connected in an external circuit (the "n" rods circuit remains open) equalization of charges carried by the holes will take place. That "p" electric current will be directed from the cold to the hot source, and it will die out after the equalization of the holes concentrations of the two rods.

If a fully closed electric circuit is present, continuous motion of electric charges in both directions will take place: "n" charges will move from the cold to the hot body, and "p" charges will move in opposite direction. The process will not damp since the internal ionization in the two contact areas would take place with different intensity that would guarantee a continuous electric driving force.

A number of attempts to create thermoelectric converters or heat generators of electricity have been made since Seebeck's discovery (1823). The efficiency of those converters is assessed via a conversion coefficient reading

$$\eta_{th} = \eta_{Carnot} * \eta_{rel},$$

where

- $\eta_{Carnot} = (1 - (T_2/T_1))$ Carnot coefficient
- $\eta_{rel} = M - 1/M + (T_2/T_1)$ coefficient of relative efficiency resulting from the irreversible losses
- $M = \sqrt{1 + 0.57 Z(T_1 - T_2)}$.

For a long time, the coefficient of relative efficiency η_{rel} was too low ($\eta_{rel} < \approx <0.002$). Hence, those thermoelectric converters were only used as temperature sensors in measuring instruments. The reason for their low efficiency was the large internal Joule work done during electric current flow through the converter and its contact "n–p" area.

---

* Here, the conductors are prepared from constantan (a copper–nickel alloy invented by E. Weston in 1887), where $\lambda = 19.5$ W/mK, and it is a semiconductor of "p" type.

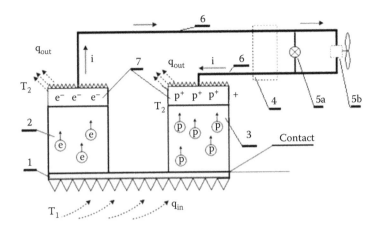

FIGURE 3.6
Thermoelectric energy generator: 1—recuperator (hot source); 2—semiconductor type "n"; 3—semiconductor type "p"; 4—inverter; 5a and 5b—users; 6—conductor of electricity; 7—surfaces releasing heat to the old source (radiators).

Recently, thanks to the pioneering works of A.F. Jeffe, an almost two order of magnitude increase of the coefficient of relative efficiency has been attained, where $\eta_{rel} \approx 0.28$.

The total theoretical conversion coefficient η_{th} of the instrument for $T_2/T_1 = 0.79$ (corresponding to $T_r = 107°C$ and $T_c = 27°C$) will be 6%.* Yet, if the thermal–electric convertor operates at a higher temperature difference, η_{th} may reach values of 14%–20%.

Those impressive results were obtained by optimization of a profit function of the form $Z = \alpha^2 \Omega / \lambda$, where Ω and λ are the coefficients of material electrical and thermal conductivity, and α is an exponential function. As a result of a number of theoretical and experimental studies, it was proved that most appropriate are alloys able to provide large specific concentration of electrons in the contact "n–p" area ($N_{eEl} > 10^{18} \sim 10^{20}$ electrons/sm^3)—see in Figure 3.5.

A typical thermoelectric energy generator employing the Seebeck effect is built according to the scheme in Figure 3.6. It is used for the recuperation of waste heat of combustion products, exhausted by heat engines, CCs, electric transformers, cooking stoves, illuminating systems, foul air released by air conditioners etc. The heat is harnessed in the generation of electricity for private needs. Recuperator 1 in Figure 3.6 is located in an environment where waste heat (*hot source* (HS)) is released, receiving heat flux with intensity q_{in}.

The remote ends of the two semiconductors contact the cold zone (*cold source* (CS)) via radiators 7 releasing heat flux q_{out}. The resulting electric driving force generates electric current running through the electric circuit consisting of cables 6 and 6a, inverter 4 and various switched-on users–LED 5a, ventilators 5b, etc.

The second interesting application of thermoelectric convertors concerns the operation of the inverse process discovered by Peltier/Thompson, where electrical energy is converted into heat and frost. Really, when electric current flows in the electric circuit of the convertor, a temperature difference occurs between the radiator and the recuperator which can be used for the air conditioning of small volumes or individual workplaces.

---

* The mean annual efficiency of an amorphous photovoltaic cell is about 8%.

3.1.3 Thermoelectric Technology for Oxidation Control in FCs (Technology of Schonbein/Grove)

FCs are power equipment where the controlled oxidation of gaseous fuel (hydrogen, natural gas, methane, syngas, propane, hydrocarbons, etc.) is carried out. As a result of the process running in a catalytic environment, electric energy* is also generated besides heat and water (which is newly generated chemical product). Currently, a large variety of FCs exists, and part of them are manufactured and marketed on a large scale. Despite the different scheme of their classification, the thermal processes use a classification that considers the value of the temperature of reaction operation. According to that classification, the FCs are divided into two groups

- "Low-temperature" FCs where the operational temperature is lower than 250°C
- "High-temperature" FCs

In what follows, we shall consider that bothof these classifications.

The group of "low-temperature" FCs usually includes: cells with a polymer membranes (PEM), alkaline fuel cells (AFC), and phosphorus-acidic cells (PAFC). The gaseous fuel used (often hydrogen or natural gas) is injected at high pressure through the intake aperture 1 in Figure 3.7. Meanwhile, pure atmospheric air (containing 21% oxygen) is injected as an oxidizer through the second intake aperture 3. The following exothermal chemical reaction runs in a classical low-temperature FC, where pure oxygen is used

$$4e^- + 4H^+ + O_2(g) \Rightarrow 2H_2O + Q, \tag{3.1}$$

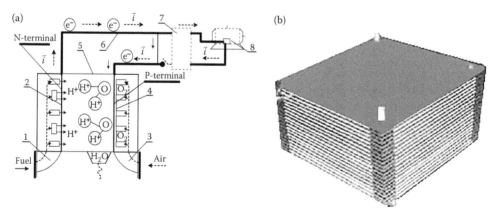

FIGURE 3.7
FC components: (a) scheme of a FC; (b) inside view of SOFC: 1—fuel supply line; 2—anode (N-terminal); 3—air duct; 4—cathode (P-terminal); 5—electrolyte (conducting medium); 6—external conductor; 7—accumulator (or inverter); 8—lectric motor.

---

* FC are classical co-generators: the chemical work done under oxidation (burning) is converted into electric and thermal work. The real conversion coefficient of modern combustion is equal to 75%–80% under a co-generation regime.

while free electrons are also produced and accumulated at one terminal, named the negative terminal. They are controllably directed by Coulomb's* force through the external copper conductor (see Figure 3.7) to end users as an electrical current.

Fuel and oxidizer are separated from one another by an electrolyte (solid or liquid electroconducting medium). Zirconium oxide (porous ceramics), hard polymers of alkaline/acidic solutions are successfully used as electrolytes.

When a solid electrolyte 5 is used in a FC, its facial surfaces 2 and 4 adjacent to the fuel feeding pipes 1 and 3 are coated by a catalyst and transformed into electrical charges terminals (electrodes). When the electrolyte is liquid (phosphorus acid or alkaline solutions, for example), the two fuel feeding pipes serve as electrodes, and they are immersed in the electrolyte, being previously coated by a catalyst.

The FC is a typical representative of an open and noninsulated TDS with three degrees of freedom doing chemical, thermal, and electric work. The process runs as follows. The compound necessary for the chemical reaction is fed continuously at pressure through the fuel and air feed pipes 1 and 3. The hydrogen molecules dissociate into anions $H^{(+)}$ and electrons $e^{(-)}$ (the catalyst facilitates migration of the hydrogen ions into the electrolyte, only, and displays large electric resistance with respect to the free electrons). Hence, hydrogen anions are subjected to potential and electric repulsive forces, and they start moving within the electrolyte. Meanwhile, the oxygen atoms join themselves as free electrons at the P-terminal and are converted in negative ions. This produces a shortage of electrons at the positive terminal (electrode)—see Figure 3.7a. Unless electron deficiency is overcome, the described processes will slow down. If a copper conductor is connected between the P- and N-terminals, the electrons shortage will be removed by the flow of electric current between the two electrodes. The electrons released by the hydrogen atoms continue their motion along an external circuit as electric current which is supplied to the connected electric appliances 7 and 8—Figure 3.7 to do the needed work. The final electron destination is the cathode 4 where electrons combine with the newly arrived oxygen molecules to transform them into cations.

The hydrogen ions are directed toward the oxygen ions (located at the other side of the electrolyte), for which they have chemical affinity. Carriers of the electric current in the electrolyte of a low-temperature FC are the hydrogen anions $H^{(+)}$. They diffuse through the electrolyte reaching the cathode where they combine with the "waiting" oxygen cations $-O^{(-)}$. The internal electric circuit closes when two hydrogen anions combine with one oxygen cation to form a molecule of water. The chemical reaction (3.1) describes the process. It is continuous when the FC is continuously supplied with fuel and oxygen, while products of oxidation and co-generation are used. Note that oxidation is an exothermal reaction producing hot water or steam, while co-generation produces electricity.

While low-temperature FCs produce hot water, mostly, the high temperature ones produce hot water or superheated steam, since their operational temperature reaches values of up to 950°C. The group of high-temperature FCs includes those with solid-oxide electrolytes (SOFC) and electrolytes composed of molten carbonate FC (MCFC). Both FC types are not sensitive to impurities and CO that might be present in the fuel.[†] Hence, they successfully operate using natural or synthetic gas, methane, and propane. If fuel contains an impurity, it poisons (contaminates) catalysts (Pl, graphene, cobalt and cobalt oxide, etc.) which cease to be effective.

---

* French physicist C. A. de Coulomb (1736–1806).
† In SOFC electricity carriers are the oxygen captions $O^{(-)}$, while in MCFC—the anions $CO^{(+)}$.

The maximal electric driving force of a FC single module, generated as a potential difference between the electrodes, is small (about 1.2 V). Yet if 10 modules are assembled in a serial group (see Figure 3.7), an output voltage is generated. It is commensurable with that of the lead–zinc batteries. The final output voltage of the FC used in systems of energy supply known as "energy servers" is 400 V, while the nominal consumed power is 100 kW.

The theoretical conversion coefficient of a FC is calculated as

$$\eta_{th} = \frac{L_{Total}}{(\Delta H^0)} = \frac{n \cdot N_e \cdot e_0 \cdot \Delta V_{max}}{\Delta H^0} = 0.85,$$

or $\eta_{th} = 85\%$. The following notations are used:

- L_{Total}—total work done in a TDS (the sum of electrical and thermal engines done per unit mass oxidizer [per 1 kmol oxygen]). It is calculated as a difference between the values of the Gibbs free potential ($L_{Total} = 56.6899$ kcal/gm-mole per unit mass)
- N_o—Avogadro number (6.025×10^{26})
- n is the number of electrons absorbed by the oxygen molecules to produce 1 gm-mole H_2O
- $e_0 = 1.062 \times 10^{-19}$ eV—energy of a single electron
- ΔH^0 is fuel enthalpy (the quantity of energy released under controlled fuel oxidation—$\Delta H^0 = 68.3174$ kcal/gm-mole for hydrogen)
- $\Delta V_{MAX} = L_{Total}/n \cdot N_0 e_0 = 1.23\,V$

The real conversion coefficient of a FC during cogeneration, however, rarely exceeds 75%.

Low-temperature FCs are efficient and durable under steady applications. Yet, one should avoid the use of hydrogen produced by natural gas reforming, since it contains carbon monoxide which contaminates the catalysts. It is recommended that hydrogen be used that is produced by electrolysis. Reformed alcohol, methanol, or ethanol can also be used.[*]

Phosphoric-acid FCs having power within the range 100–200 kW[†] are used for building applications. Under co-generation FCs their thermal to electrical outputs are in proportion as 40/36. Their conversion coefficient attains values of up to 76%. A typical FC arrangement includes a preprocessor (reformer), which produces hydrogen from the fuel, a stack where electrochemical processes are in operation and an inverter (power conditioner) which transforms the DC current into AC current. Most FCs include also an interconnector (a device that synchronizes the parameters of the output current with those of the grid—ASHRAE 2001; ASME Standard PTC 50).

FCs with molten carbonates generate power of 100 kW–10 MW, while those using solid oxides 100–200 kW. Their operational temperature is 650°C, and cogeneration efficiency reaches 70%. They are not "choosey" with respect to fuel purity, and they become more and more popular in thermoelectric projects.

---

[*] There are FCs which operate directly with alcohol (DAFC), but their power is low.

[†] Individual PAFC generate power of up to 10 MW. In principle, their start is difficult (the electrolyte should be heated at every start since the phosphorus acid is solid below 40°C). They are very suitable for steady operational models.

3.2 Indirect Thermoelectric Technologies

The group of indirect technologies includes those capable of converting the generated heat into kinetic energy of mechanical bodies. Then, as shown in Figure 3.8, the mechanical energy is converted into electricity via electromagnetic induction.[*]

Figure 3.8 shows that there are four technologies of heat generation ordered with respect to their market share—uncontrolled oxidation (combustion), fission of the atomic nucleus, solar and geothermal heating. Following the above order, we shall discuss in what follows the two most popular thermoelectric technologies:

- Rankine's steam-turbine technology of
- Brayton's gas-turbine technology of

The two technologies are similar to each other.[†] The difference between them is insignificant at first glance and concerns the type of the used working body, that is, Rankine technology uses fluids (water, chlorofluorocarbons, lithium–bromide mixes, etc.) which undergo phase transformations, while Brayton technology uses combustion gases exhausted during the burning of the primary energy carrier. In fact, there are significant differences concerning technological equipment and efficiency

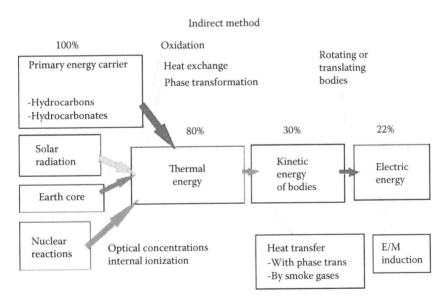

FIGURE 3.8
Indirect thermoelectric technology.

---

[*] M. Faraday (1791–1867) is considered to be the man who discovered electromagnetic induction in 1831, after having designed and published his single-pole generator (1832–1833). Yet, four years earlier (in 1827) the Hungarian engineer Anyos Jedlik (1800–1895) built a working electric generator but it was not officially recognized untill 1850.

[†] They belong to the class of externally fired engine technologies where the combustion devices are physically insulated from the power devices—steam and gas turbines in this case.

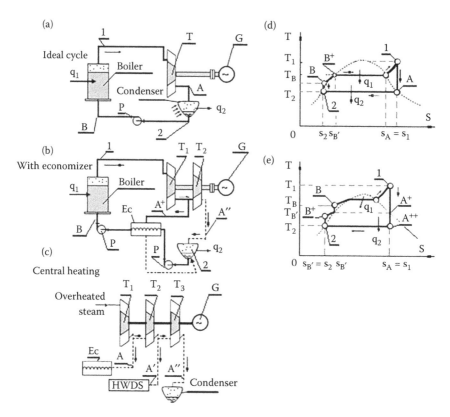

FIGURE 3.9
Steam-turbine technology of Rankine: T—steam turbine; G—electrical generator; Ec—economizer; HWDS—heat exchanger of the system for hot water domestic supply; P—feed pump (a) ideal cycle of Rankine; (b) recuperation cycle along the line of the steam processed via the economizer; (c) as in (b), but with heat exchanger for domestic hot water supply; (d) diagram of the ideal process; (e) diagram of the recuperation process (*Points: 1, A, 2, and B in the schemes correspond to those in the state diagram of the working fluid*).

1. Rankine technology realizes energy transfer between the heat generator (boiler) and the thermal engine (steam turbine) using smaller amounts of fluid mass. The energy is "encoded" in a "liquid–gaseous phase" transformation (i.e., transformation of a liquid into superheated steam as shown in Figure 3.9. To close the cycle however after steam exhaust, one needs to perform inverse steam conversion (forced condensation) in a special device called condenser. In Brayton technology, heat is transferred by combustion products directly to the thermal engine (gas turbine). To transfer energy amount comparable to that of the Rankine technology, one should use more amounts of flue gases. This decreases the technology efficiency to a certain extent. Yet, the necessity of condensing the working fluid falls away and the inevitable heat losses along the whole technological circuit are avoided—note that they are significant in pure electrification technologies, amounting to 50%–60%. The condenser as a component of the system is also removed.

2. The second essential difference between the two technologies is that between their temperature ranges of operation. In Rankine technology, the temperature of the superheated steam is $T_n \approx 550$–600 K, while that of the condenser is about 353–408 K. This yields a relatively low conversion coefficient

$$\eta_{th} < \eta_{Carnot} = 1 - \frac{353}{600} = 0.588 \Rightarrow 58.8\%.$$

The real conversion coefficient of this technology does not exceed 38%.

As for Brayton technology, energy transfer takes place at significantly larger temperature differences. For instance, if the temperature of the combustion products is $T_1 = 1573$ к (see Figure 2.25), while that of the exhaust gases drops to the temperature of the surroundings ($T_2 = 300$ K), the efficiency of the thermoelectric technology increases significantly, since $\eta_{Carnot} \sim 81\%$, and the real value of the conversion coefficients approaches 65%.

A disadvantage of Brayton technology is the necessity to use only high-temperature resistant materials for the manufacture of the technological equipment (molybdenum and nickel steels, titanium alloys, ceramics, and metal powders). This significantly increases the amount of the initial investment. On the other hand, it is known that if there is a market that is hungry for the products of modern metal science, product prices go down significantly. There are a number of examples in this respect, and the development of the energy sector in the next quarter century seems promising. That optimism is verified by the invention of new and more efficient materials (see for instance the products of nanotechnology).

3.2.1 Steam-Turbine Technology of J. Rankine

Historically, Rankine[*] technology has been a special service to modern electrification,[†] and there are two technology types. The first one is that of electricity generation without recuperation of the waste heat (Figure 3.9a). The second one is co-generational steam-turbine technology for the simultaneous generation of electricity and heat in the so-called "central-heating" or public thermal electric power stations (Figure 3.9b).

Regarding the first type, heat q_1 generated by burning of primary energy carriers—fuel (coal, wood, oil derivatives, etc.), is supplied to the working fluid right in the CC. There, the working body changes its aggregate state and (usually water) evaporates isobarically along the line "B–B$_1$–1'–1." Thus, the first stage of the Rankine cycle (isobaric expansion) is completed at point "1" by preheating of the working fluid into superheated steam with temperature T_1—see Figure 3.9d).

To start the second stage (adiabatic expansion) the working fluid is expired into the steam turbine where it releases its internal energy to do mechanical work: $l_{M1-A} = -u_{1-A} = C_v(T_1 - T_A)$.

It overcomes the friction moments in the turbine and generator bearings, as well as the electric resistance moments. Under adiabatic expansion, the temperature of the working fluid drops from T_1 to T_A.

The above equality shows that the mechanical work done by the steam turbine depends on the difference between temperatures T_1 and T_A, and the larger the temperature difference the more efficient the process. When the steam used in the turbine attains temperature T_A, it is introduced into the condenser where it is again condensed under isobaric conditions ($p_A = const$) releasing heat to the cold surroundings (the atmospheric air). The temperature of the working fluid remains constant $T_2 = T_A$. Thus, completing the third stage

---

[*] A Scottish engineer, William John Macquorn Rankine, (1820–1872) who created in 1859 the steam engine theory, describing the engine operation via a state diagram. He introduced a number of thermodynamic terms.

[†] Due to the low efficiency (~20%), this technology consumes enormous quantity of primary energy carriers. It is the largest emitter of greenhouse gases yielding global warming.

of the Rankine cycle, the fluid returns to its initial state—liquid. Heat released during condensation dissipates in the atmosphere (used as a cold body). This is a serious disadvantage of the technology.

The fourth (closing) stage of the Rankine cycle is adiabatic pumping from point 2 to point B (pressure rises from p_2 to p_B), where the feed pump pushes the working fluid in the boiler for the new evaporation process and cycle repetition. As seen in Figure 3.9a, the steam turbine is coupled with an electric generator where the kinetic energy of rotation of the working wheel and the generator rotor is *converted into electrical energy via electric induction*.

Rankine technology is *energetically* imperfect, since it "suffers" huge energy losses in the condenser (55%) and during the release of hot flue gases (12%). However, it was improved in the course of more than a century.* Initially, heat of the consumed steam (after the first section of the turbine—Figure 3.9b) was used by the economizer of the steam boiler.

Next, a heat exchanger was incorporated as a heater which increased the efficiency of the steam-turbine technology.

Steam needed for the operation of the heat exchanger is deflected after the second section of the steam turbine (see Figure 3.9c). Thus, energy co-generation is possible. In installations designed following such a scheme, about 50% of the primary energy is spent for heat generation for households, 12% is spent for the generation of electricity (co-generation), and the losses in the condenser are 20%, only. The total regeneration coefficient of the Rankine steam-turbine technology may attain 60%–65% (this is its proven maximal efficiency).

There are three more versions of the Rankine steam-turbine technology, in addition to the one already described. They are based on the use of solar radiation, Earth core energy, and the energy of thermonuclear reactions. We shall describe them in what follows, since they outline new prospects of effective development of the Rankine cycle.

Figure 3.10 shows a scheme of a solar thermal electric power station operating with Rankine technology. Heat is generated using the "optical concentration of solar radiation,"† which reaches high density of solar energy flux directed to the solar receiver (position 3a in Figure 3.10a). It exceeds by 30–40 times the nominal density.

The working medium of the first loop (water solutions of sodium and potassium salts or high carbohydrates—pentane, sextane, etc.) is heated to 640 K passing through the solar field 1, without changing its physical state. All receivers of the solar field are linked in parallel. The heated fluid passes through the steam generator 4 and storage 5 (see Figure 3.10), giving its energy to the heat carrier of the secondary loop. Then, undergoing the pressure increase by the circulation pump 6, it enters the pipe system of the solar field to be heated again and repeat the cycle. The second loop consists of the steam generator 4, steam turbine 8, condenser 10, circulation pump 7, and accumulator 5, and the working fluid is water steam.

During the operation the processes are similar to those found in the classical steam-turbine technology—see Figure 3.9. The difference is in that the classical boiler (steam boiler) is replaced by a steam generator 4, which is a highly efficient heat exchanger of the "water/steam" type. Another difference is that instead of an economizer, the heating exchange coils of the accumulator 5 are used here. Following the requirements of Rankine technology, those formal differences do not affect process efficiency. The incorporated electric generator 9 converts steam potential energy (at temperature 450–500 K) into electricity via

---

* The first power stations operating with gas turbines were built in the mid 20th century.
† Known also as the solar thermal method, CSP (concentrated solar power) or TSS (trough solar system).

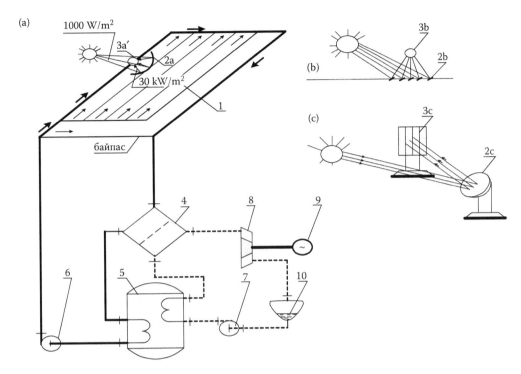

FIGURE 3.10
Solar thermal technology for the generation of electricity, arrangement of the technological equipment of the station. (a) parabolic reflectors; (b) linear Fresnel reflectors; (c) Heliostatic tower (3c) with tracing reflectors (2c): 1—solar absorber field; 2a—parabolic reflectors; 2b—reflector bands; 2c—reflector mirrors; 3a—movable solar receiver; 3b—stationary solar receiver; 3c—reversible solar receiver "Heliostatic tower"; 4—steam generator; 5—thermal storage and exchanger; 6—circulation pumps; 7—feed pump; 8—steam turbine; 9—electric generator; 10—condenser.

turbine rotation. The conversion efficiency of Rankine solar thermal technology can range up to nearly 30% (usually exceeding the efficiency of the photovoltaic solar technologies by good measure) and outruns the conversion coefficient of PV systems.[*] Considering the 2008 cost of 1 kWh power generated by solar concentrators cost ($ 0.20), as well as the prospects of cost decrease during the next 10 years, the technology may gain ground in Mediterranean countries, including Europe.

The second important application of the Rankine cycle[†] is the steam-turbine technology for generation of electricity. It utilizes the geothermal energy of the Earth core[‡] using the method "hot dry rocks—HDR"—see Figure 3.11.

A refrigerating agent is the working fluid in this case, boiling at low temperature (water–ammonia solution, for instance). The pressurizing pump 8 injects deep into the ground the liquid (body) to reach the so-called "dry hot rocks" 9. There, it evaporates and the superheated steam is sucked by the compressor 10 and pressurized into the

---

[*] At present, about 3000 MW are installed worldwide, mainly in the USA—California and Nevada; about 100 MW in the EU—Grenada, Spain and 140 MW in the Negev desert. Investments amount to $7.14 × 10⁶/MW.
[†] This technology is also known as *"The cycle of A. Kalina."*
[‡] As is known from geology, the temperature of the Earth layer grows by 3°C per each 100 m depth. The so-called in-depth drilling of 3–5 km reaches the "hot" rocks, whose temperature is about 100–200°C (as in Basel, Switzerland, for instance).

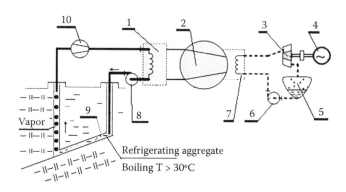

FIGURE 3.11
Geothermal technology (dry hot rocks) for the generation of electricity: 1—evaporator of the heat pump; 2—heat pump; 3—steam turbine; 4—electric generator; 5—condenser; 6—feed pump; 7—condenser of the heat pump; 8—pressurizing pump; 9—HDR; 10—sucking compressor.

reservoir-evaporator 1. Devices 1, 8, 9, and 10 participate in the primary loop of the geo-thermal installation. That loop takes heat out of the hot rocks and transports it to the heat pump 2. Its operation is hydro dynamically independent of that of the rest two loops. Yet, it determines their energy efficiency.

The heat pump 2 forms the *second loop* which is also hydro dynamically independent. The pump takes the energy of phase transition of the working fluid via its evaporator 1, where the working body condenses and liquefies for secondary usage. The heat pump is calculated such as to operate in *"steam/steam"* regime. It increases the temperature of the Rankine loop's working body in condenser 7 of the HP and it is evaporated. The steam boiler here is replaced by the condenser section 7 of the heat pump producing steam for the operation of turbine 3. Since, water successfully is evaporated in device 7, thermal energy is transferred to the steam turbine, which is coupled with an electrical alternator in the *third loop* of the system. Used steam is condensed in the condenser 5 (a compo-nent of the Rankine equipment), while the feed pump 6 pressurizes it back into the heat exchanging tubes of the HP condenser 7 to be again evaporated, thus maintaining the cycle performance.

The total conversion coefficient[*] of geothermal stations reaches 95% and electricity cost is about $0.15/kWh (seldom—$0.05/kWh). At present, it is twice higher[†] than that of wind electric power stations and only three times higher than that of electricity generated by coal-burning thermoelectric power stations. The technology is absolutely clean and renew-able, leaving minimal traces in the environment. However, its greatest advantage is that the needed heat is freely and readily available—at a depth of 3 km, only.

Last but not least, Rankine technology is applied in nuclear power stations (NPSs). The main problems of the nuclear technologies are how to guarantee environmental protection, safe and healthy working conditions, and reliable and safe facility operations. So-called "multi loop" systems are used for the sake of safety, where the risk of outflow of radioac-tive materials is reduced. Figure 3.12 shows two schemes of arranging the technological equipment—with *two and three loops*. Note that the increase of the number of internal loops (from 1 to 2) yields significant increase of the reliability and thermal inertia of the system.

---

[*] The Rankine technology energy losses are balanced by the heat pump operation.
[†] Drills are rather expensive and risky—the working fluid may be lost as was in Bad Vram, Stuttgart, for instance. The elimination of such problem costs about 2–4 million €. The whole investment amounts to 6.5 million €.

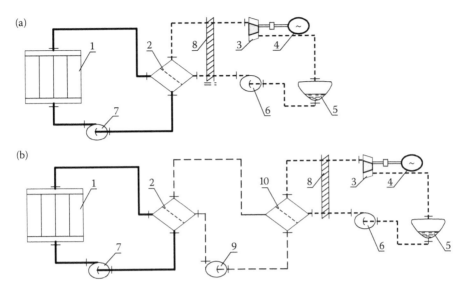

FIGURE 3.12
Steam-turbine NPS–multi loop schemes of NPS. (a) Two-loop; (b) Three-loop: 1—reactor; 2 and 10—steam generators; 3—steam turbine; 4—electric generator; 5—condenser; 6—feed pump; 7 and 9—circulation pumps; 8—biological protection.

Besides, all components of the internal loop, which are in direct or indirect contact with the reactor (for instance, circulation pumps 7 and 9, steam-generator heat exchangers 2 and 10 and automation components), are inaccessible thanks to biological protection 8. It is the second important feature of this technology. In fact, reactor biological protection is guaranteed by an insulating layer of massive concrete covered by lead, steel, or aluminum sheets. Thus, it is impermeable to all radioactive products. The remaining units are the same as those of the Rankine technology. They are similarly arranged in an external loop consisting of steam turbine 3, condenser 5, and feed pump 6.

The enormous energy inertia of the nuclear reactors intensified by the thermal inertia of the other two loops makes the NPSs difficult to be centrally regulated. Hence, they are used to support the stationary season loadings of the electric supply system, only, forming the system base. All other non-nuclear energy sources can be used as either seasonal or 24-hours energy reserves. Soon, they will play an important role in the conclusion of local energy contracts after the activation of a regional (e.g., Black Sea coast) energy stock market.

Hence, despite its low efficiency, the Rankine technology of electricity generation or co-generation can be employed anywhere where phase transformation of the working fluid is to be performed. Included are atomic reactors that operate using light water, solar thermal systems, or systems for the utilization of highly potential energy waste from industrial technologies.[*]

The Rankine cycle was replaced in transport in the 20s by other engine technologies and recently—by hybrid technologies (see Section 3.3). Exceptions are only cases where nuclear technologies are applied—submarines or space rockets, for instance.

Based on those specific technologies, nuclear reactors maintain favorable chances of being used as future energy reactors and co-generators for domestic and industrial applications, while their overall dimensions would drastically decrease.

---

[*] Waste from metallurgy, building or food industry.

TABLE 3.2

List of Nuclear Reactors with Capacity Smaller than 50 MWe

No	Type	MWt/MWe	Designer	Cooling System	COP (%)
1	GEN4 Energy	70/25	Hyperrion	HPG	36
2	KLT-40S	150/35	OKBM	PWR, 4c.l.	23
3	NuScale	160/50	NuScale&F	Integral PWR	31
4	SSTAR	45/20	LL&LANL	Carbon dioxide	44
5	Toshiba 4s	30/10	Westing H.	Sodium cooling	33
6	MRX	50/30	JAERI	PWR	60
7	Unitherm	20/5	RDIPE	PWR	25
8	Shelf	28/6	NIKIET	PWR	21
9	Triga	164/64	General A.	PRW	39
10	Smart Dunedin	30/6	Dunedin ES	–	20
11	Sealer	8/3	Royal IT	Lead CR	37
12	CEFR	65/20	China	Sodium cooling	31
13	Rapid-L	5/0.2	JAERI	Molten Sodium	4

Table 3.2 gives a list of nuclear reactors with power less than 50 MWe, that are in the process of implementation or already in operation, but not for military and transport usage, thus confirming the tendency to miniaturize nuclear reactors with nuclear fission. They are part of modern heat generators operating after Rankine's or Brayton's schemes.

3.2.1.1 Tendency to Miniaturize the Overall Dimensions and Power of Nuclear Reactors

Nuclear reactors operating with "fast neutrons" of nuclear fusion, a "Tokamak" class, might be usable for energy generation, but not until the second half of the present century.

In the last 10–15 years, worldwide leading nuclear research institutes and companies designed over 30 reactors with power not exceeding 50 MWe (see column 3 in Table 3.2), which operate with "fast neutrons," yet generated by nuclear fission. The characteristic feature of most of them is that the "energy extraction" from the reactor core is carried out by phase-converting coolants following Rankine's thermodynamic cycle (with coefficients of heat-electricity conversion in the range of 4%–60%—column 6 of Table 3.2). Besides water, easily melting metals such as lead, lithium or sodium, for instance, are also used as coolants.

In fact, small nuclear reactors were used in energy generation even in the 20th century. The smallest, with power of 12 MWe, was installed in the "Bilbino" NPS, Chukotka, Russia, as far back as 40 years ago. That reactor known by its acronym EG-6 is of water-graphite type (LWGR). Another emblematic example is the reactor BN-20, operating in the town of Tueali, China, with power of 25 MWe and 65 MWt, respectively.

At present, the least powerful reactor is integrated with the nuclear co-generator Rapid-L, designed in Japan by JAERI (1999–2001), with power of 0.2 MWe and 5 MWt, respectively, followed by the reactor incorporated in the co-generator Sealer with power 3 MWe and 8 MWt designed by the Royal Institute of Technology of Stockholm.

Rapid-L is an energy co-generator (6:0.2 MWt/e), using liquid metals as coolants, such as lithium Li6 (T_{melt} = 1063°C); sodium (T_{melt} = 882°C) or potassium (T_{melt} = 757°C). Lithium is also used as a filter of neutrons, emitted in the course of the nuclear reaction.

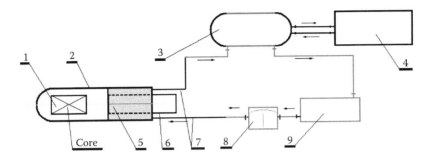

FIGURE 3.13
Scheme of the nuclear co-generator Rapid-L of JAERI with power 5.0/0.2 MWt/e: 1—integrated fuel assembly; 2—reactor; 3—power conversion segment; 4—heat exchanger (for heat supply); 5—heat exchanger (sodium heat pipe); 6—LRM, release mechanism; 7—sodium circulation lines; 8—electromagnetic pump; 9—sodium tank.

The circulation of the liquid metal within the cooling system is realized by electromagnetic pumps.

The Rapid-L reactor is in operation without human intervention for a long period of time (10 years). Uranium–nitrate mix is used as fuel, in proportions 2:3, while a revolving mechanism is currently employed to replace cassettes with discharged fuel.

Figure 3.13 shows a general scheme of the co-generator Rapid-L. It is seen that the reactor core 1 is separated from the electrogenerator 3 by a two-loop cooling system, consisting of a heat exchanger 5 and a system for the circulation of liquid sodium comprising an electromagnetic pump 8, circulation lines 7, and a tank 9.

Besides the two-loop system thus designed to cool the reactor, the high reliability of Rapid-L is alsoguaranteed by a scheme for automatic regulation, consisting of four subsystems

- LEM—for control of the liquid metal
- LRM—for automatic start of the reactor
- LEM and LIM—for current automatic control of the reactor (LEM—for feedback control and LIM—for end control)

The overall dimensions of Rapid-L are 2 m diameter and 6.5 m height, while its weight is 7700 kg. The efficiency for heat conversion into electricity is 4%, which makes the device suitable for operation in a cold climate. Initially, it was designed to operate in space, while the compact structure makes it suitable for use in domestic energetics with thermal loads multiply exceeding the electric ones. The Rapid-L co-generator is installed underground, in a shaft under the border of the building it served, while the earth layer over it serves as a biological barrier against radiation.

The second compact micro reactor included in our survey is the Swedish co-generator Sealer (see Figure 3.14). It is designed to operate with an efficiency coefficient (EC) of 31% and nominal heat and electric power of 8.0/3.0 MWt/e, respectively (see Table 3.2). A mix of uranium dichloride (UO_2) and enriched uranium (U) in proportion 4:1 is used as fuel. Cooling is carried out by a liquidized metal (lead, with melting temperature $T_{Melt} = 450°C$).

The dimensions of the "miniature" Sealer reactor (capsule diameter $\Phi_c = 2.75$ m and height—$H = 5.9$ m) enable one to incorporate it either in stationary or in transported systems. The value of the EC is 38%.

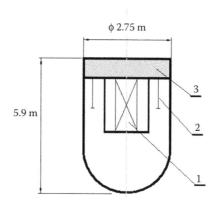

FIGURE 3.14
Scheme of the nuclear microreactor "Sealer" of the Royal Institute of Technology of Stockholm with power 8.0/3.0 MWt/e: 1—core; 2—supporters; 3—heat exchanging structure.

Nuclear devices operating with fast neutrons may be summarized as having the following tendencies:

1. Conversion into micro co-generators. Yet, the expected capacity for domestic users of 0.03–0.05 MW has not been attained

2. Pursuit of increasing the temperature of electricity generation (about 600–700°C) and use of coolants with thermal capacity larger than that of water (for instance, liquid metals such as potassium, sodium, and lead)

3. Attaining more compact overall dimensions applicable in urbanized areas. Pursuit of higher reliability and more safety

4. Increase of reactor energy efficiency during thermoelectric conversion employing more efficient thermodynamic cycles—Brayton's cycle for instance

3.2.1.2 Alchemic Future of the Thermal Technologies Used for Domestic and Industrial Needs?

Modern* interest to the secrets of alchemic "transmutation" of metals was aroused in 1989 after one or two centuries standstill, when M. Fleischmann[†] and S. Pons from the University of Utah published their experimental finds (Fleischmann and Pons, 1989). They set forth a description of the successful conversion of nickel (Ni) into copper (Cu)[‡] under relatively low temperature. The transmutation reaction was strongly exothermal which impressed researchers. The released heat exceeded several times that of the chemical reactions. The explanation of this phenomenon involved the design of a new nuclear reaction of

---

* Alchemy is an old practice originating from ancient Egypt (30–20 BC), having acquired a common context during the Late Middle Ages when attempts to "transmute" base metals into "noble" ones (particularly gold) were made. This term is used in the present book to generalize the studies in the field of low temperature nuclear synthesis (LTNS), where conducting metals (Ni, Pd, and others) are converted into chemical elements with higher molecular weight (Cu for instance) by hydrogen and deuterium absorption. These reactions are exothermal.

† M. Fleischmann (1927–2012)—British chemist, born in Karlovy Vary, Czech Republic, premature announcement of the Cold Fusion process.

‡ Nickel grid is used in the reactor, which filters a flux of deuterium (heavy nitrogen).

the atomic nuclei corresponding to copper (Cu) where the excess nuclear energy is released into the reactor core and converted into heat.

In modern physicochemistry, that process is known as cold fusion (cold nuclear synthesis [CNS]), while the devices where it takes place are called low-energy nuclear reactors (LENR). The cold fusion runs without emission of radioactive particles or radioactive co-products, since it takes place under the effect of so-called "slow neutrons."[*] Hence, it is expected that such various "alchemic transmutations" would well fall into place in future thermal technologies for domestic and industrial needs.

During the last two decades after the publication of the paper of Fleischmann and Pons, many innovation companies and single inventors were inspired and motivated by the idea of implementing the CNS in the design and building of heat generators, boilers, or water heaters. One of them was the Italian inventor Andrea Rossi (Rossi, 2015). His fluid heater is of module type with power of the individual module amounting to 10 kW, while the package described in Cook and Rossi (2015), consists of 100 modules. It is under a 10-month acceptance test since the beginning of 2015, displaying the efficiency in the range $6 < EC < 30$.[†]

Figure 3.15a shows a scheme of one of the modules of eCat, which is shaped as a rectangular arm with dimensions $0.3 \times 0.3 \times 0.08$ m consisting of a horizontal and a vertical section. The heater is periodically charged (every 6 months) with fuel consisting of 50% nickel, 20% lithium, and 30% lithium–aluminum hydrate. In addition, the module is charged with hydrogen and cooled with water. Water intake and exhaust temperatures and water outflow are measured (see Figure 3.15a). Rossi (2015) explains the high efficiency of the device via the following mechanism: "…A proton from a hydrogen atom enters, by the quantum tunneling effect, into a nucleus of Li-7 (i.e., a lithium nucleus of atomic weight 7), forming a nucleus of Be-8 (i.e., a beryllium nucleus of atomic weight 8), which then decays in a few seconds into two alpha particles (helium nuclei)… ." The, excess energy of fracture of the interatomic bonds is released during the process flowing into the reactor. An axonometric view of a module package with power of 1 MWt is shown to the right of the scheme of eCat fluid heater—Figure 3.15b). It consists of 100 parallel-linked components undergoing 1-year trial operation (Cook and Rossi, 2015). Other inventors as for instance the Russian physicist A. Parkhomov (2014) designed and built analogs of Rossi's device but operating under different temperatures.

Generalizing the studies on the development of low-temperature nuclear reactors, we should note that pursuant to NASA studies the global energy consumption of modern civilization can be covered by only 1% of nickel produced worldwide.[‡]

3.2.2 Brayton Gas-Turbine Technology

This technology engages the direct Brayton cycle to eliminate the disadvantages of the Rankine cycle using the energy of the direct combustion of hydrocarbon fuels more effectively. This is the third modern application of the Brayton cycle, after gas turbine and jet

---

[*] With power lower than 1.0 eV.

[†] Due to Rossi's explicit denial to reveal the internal structure of the ECat heater, doubts about the correct measurement of the efficiency coefficient still remain.

[‡] A LENR technology, different from atom fission and fusion has been developed by NASA. The energy is generated at the interface between the nickel lattice and hydrogen atoms under the influence of an external force field oscillating at a frequency in the range $5 < f < 30$ THz. This action ionizes hydrogen atoms, turning them into protons and stimulates a flow of slow (with low energy) neutrons. The nickel absorbs those neutrons instantly and goes into an unstable form. It regains its stability, becoming copper. The reaction is exothermic.

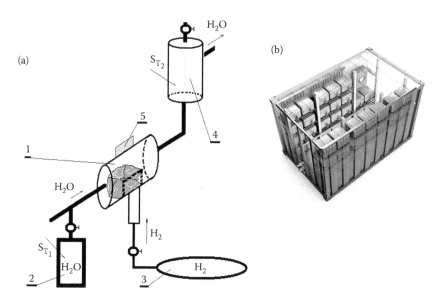

FIGURE 3.15
Fluid heater eCat of A. Rossi: (a) structure of heating module with power 10 kWt; (b) heating package with power 1 MWt, composed by 100 modules: 1—reactor; 2—water tank; 3—hydrogen tank; 4—vertical pipe; 5—control block.

engines. Formally, the technology arrangement was similar to that of the steam turbine.[*] Heat-converting equipment consists of water phase transformations including condenser, condensing reservoir, and feed pump, since fuel gases released in the CC operate as a working fluid. Figure 3.16a shows an arrangement of the devices realizing the gas-turbine cycle of Brayton, as well as the p–v diagram of the respective processes. The new element of the technological scheme is the turbo-compressor C for air compression performed prior to combustion (could be one stage or multistage of compression). The processes are

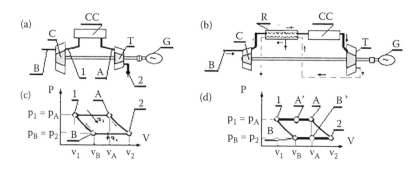

FIGURE 3.16
Gas-turbine technology of Brayton: C—compressor; CC—combustion heat exchanger. (a) Ideal cycle without utilization; (b) cycle including heating of the combustion products, utilization; (c) p–V diagram of the ideal cycle without utilization; and (d) p–V diagram of the cycle with the combustion products utilization.

---

[*] The first gas-turbine installation was designed by B. Bowery in 1939 in Noshal, Switzerland. The device power was 4 MW (NYSEG-Microturbine Demoot), and the conversion coefficient amounted to 18% only (its operational temperature however was within limits 537/277°C).

similar to those running in the gas-turbine engine (see Section 2.3), and the efficiency of the gas-turbine cycle is assessed as

$$\eta_{ett} = \frac{L_C}{q_1}.$$

The first installations[*] of gas turbine had a separate CC from the remainder engine and operated at relatively low temperature, with small temperature difference and resulting low efficiency. The technology was improved[†] in the 1970s in three ways

- Increase of the temperature of the intake gases (1700 K) and increase of the conversion efficiency with co-generation—up to 80%–85%
- Arrangement of all components (compressor, gas turbine, and CC) in a common casing or block—see Figure 3.16 (as is in the gas-turbine engine)
- Expanding the range of available power (at present, the minimal power of gas microturbines is 3–6 kWe, while the maximal power approaches 600 MWe)

High-temperature resistant nickel and cobalt steels are used for gas microturbine manufacture. Nowadays, so-called "single crystal" technologies are also applied and the thermal schemes have improved. For instance, heat of the flue gases is recuperated and used for additional heating of the fresh but already compressed air, prior to its injection into the CC (Figure 3.16b). Other even more complicated schemes include additional combustion products heated by injection of additional fuel to increase gas temperature at the entrance of the turbine second section—see Figure 3.16c.

To increase efficiency, the air compressor should also be improved. Since the compressor is combined to the gas turbine (thermal engine), intercooling of the compressed air is performed (between section 1 and section 2) in order to decrease the operational power (see Section 2.5.3 discussing isothermal compression).

In conclusion, note that the gas-turbine technology for co-generation of electrical and thermal energy is applicable not only in industries where large power is needed but also in individual buildings (apartments, houses, etc.).

This is an essential advantage. Moreover, gas microturbine generators for individual use are compact and have dimensions of a washing machine or a refrigerator (see Figures 3.17 and 3.18). They are easily installed, their price is affordable, and the co-generation conversion coefficient is significant (over 80%). Yet, the availability of gas supply is operational *condition sine qua non* for those devices.

3.3 Employment of Thermoelectric and Co-Generation Technologies in the Design of Vehicle Hybrid Gears

Historically, the development of thermal, thermomechanical, and thermoelectric technologies stimulated and motivated innovations in transport. For instance, only 15 years after

---

[*] For instance, those in Noshal or those of Mitsubishi Heavy Industries (1940).
[†] Over 15,000 patent applications are registered.

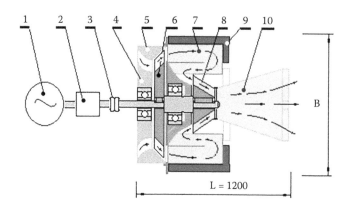

FIGURE 3.17
View of a micro gas turbine aggregate: 1—electrical generator; 2—output shaft; 3—disc clutch; 4—housing with bearing box; 5—air intake; 6—compressor; 7—CC; 8—gas turbine; 9—water shirt; 10—exhaust of combustion products.

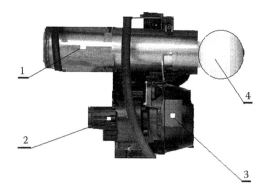

FIGURE 3.18
Micro gas co-generator: 1—heater; 2—micro turbine; 3—electrical generation output; 4—exchanger.

James Watt designed the steam thermal engine, it was mounted on the first steam car,[*] locomotive,[†] and ship.[‡] Similar attempts were made in aviation, too, but the idea was not developed because of the lack of appropriate materials. The dependence of technology effectiveness on this interrelation was confirmed in time by the invention of internal combustion engines.

Today, however, prices of primary energy carriers are rising, and public control over hydrogen and nitrogen oxides emissions of car engines has intensified. Hence, new types of driving systems (so-called "hybrid drives") are offered on the market, combining two types of power engines—thermal and electric. The task is to increase the efficiency of all land vehicles.

Hybrid drives have an almost age-old history. The pioneer in the field was the young car designer and manufacturer Ferdinand Porsche. In 1898 he designed and sold about 300 pieces of his famous hybrid car popular as the "Lohner–Porsche" model. Hybrid cars

---

[*] Invented in 1784 by William Murdoch (1754–1839).
[†] Invented in 1800 by Richard Trevithic (1792–1882) and manufactured in 1803.
[‡] Patented in 1736 by Jonathan Hulls, built in 1763 by William Henry.

became a real market hit during the last century as a result of the continuous rise of fuel prices and the "rebate incentive programs" of the EU and the USA. The relatively low cost and good reliability made them attractive for millions of customers all over the world (almost 5.2×10^6 hybrid cars were sold in 2015 year). While hybrid cars engage more and more markets and gain more prestige, the market share of their "senior" brothers—hybrid locomotives, has shown very slow rate in the last 4–5 years. Yet, the present decade awaits their market "explosion." These expectations are confirmed by the investment activity of the major companies in locomotive design, such as General Electric Transportation, Alstom, Siemens, Vosslon Loco, etc. They all have production capacity for basic hybrid components (high density accumulators, powerful inventors, highly efficient motor-generator aggregates, and reversible electric motors). The fact that the Canadian company Railpower, specializing in the manufacture of hybrid locomotives known as Green Goat, sold 32 pieces in 2005 (14 pieces just in the last quarter) and tripled their sales in 2006, is in support of the growing interest in those machines. According to the Pike Research report (Hurst, 2010) the hybrid locomotive could achieve a CAGR[*] of 19.4%, with annual unit sales up to 109 locomotives by 2020.

The manufacture of hybrid locomotives is considered to have started as far back as 1917 when *Hermann Lemp*'s prototype of called *AGEIR* was assembled by General Electric (patented in 1913), but it was only a sample (see Figure 3.19). Since the first machine, several dozen types of hybrid locomotives were designed in North America and Europe during the last century.[†]

As for transport operators–users of hybrid locomotives (with two engines), the introduction of these machines is a precondition for the economy of a significant amount of energy, avoidance of legal fines for carbon dioxide pollution, and improvement of the quality of transport. Economy of the primary energy carrier in hybrid drives is realized by recuperation of the extra power of thermal engines operating under nonsteady regimes that arise on hilly railway terrains (during train ascent or descent, at train pull up or departure, under engine floating, etc.). These are railway sections with speed restrictions depending on the number of train stations, railroad length, train type (passenger, fast, express, or freight), etc.

As is known, each speed change by 1 km/h (within the range $50 \leq w \leq 100$ km/h) yields energy overconsumption varying by $\pm 1\%$/km s^{-1}. An impressive amount of primary fuel is lost in engine floating and at train pull up (about 1000 t/year when idling and 200 t/year for starting diesel fuel). A single rolling locomotive overconsumes significant amounts of primary energy carrier, and power estimation is about 1.4×10^6 kWh/year (1 kg/km diesel fuel and 5 kWh/km electric power). Last but not least, the large over consumption of fuel is due to the inability of the crew to economically drive the locomotive on hilly terrain (this is the human factor as a regulator of the engine power).

The idea of the hybrid drive is based on effective usage of the primary energy carriers (oil derivatives, biodiesel, ethanol, methanol, methane, natural gas or syngas, hydrogen,

---

[*] Compound annual unit growth rate.

[†] In 1920, *Baldwin Locomotive Works* made a prototype of a "special" diesel-electric locomotive using electric components manufactured by Westinghouse Electric Company. The Kaufman Act prohibiting the use of steam locomotives was approved in the USA in 1923. This marked the end of the era of steam engines and the start of the era of diesel and diesel-electric locomotives. In 1929, the transport company *Canadian National Railway* became the first transport operator to use two diesel-electric locomotives manufactured by Westinghouse. In 1930, General Electric manufactured several small hybrid engines (the famous "*44-tonner*" hybrids) which were adopted by *Burlington Railroad* and *Union Pacific* in 1940 for passenger transport. Later, Southern Pacific was equipped with the two-valence (diesel-electric) locomotive EMD565 with power 4853 kW(6600 hp), as well as other improved models. The vehicles rolled along the Milwaukee Road (Santa Fe, California) till 1969–1970.

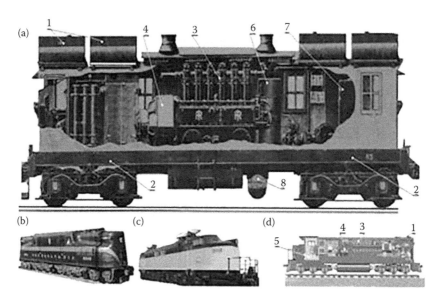

FIGURE 3.19
Evolution of hybrid locomotives: (a) prototype of the hybrid locomotive *AGEIR* of Hermann Lemp; (b) FT demonstration model #103 of general motors; (c) Germany's Deutsche Bahn AG; (d) SNCF Class B 81500 electro-diesel. 1—radiators; 2—electric motor; 3—thermal engine; 4—electric generator; 5—crew cabin; 6—fuel reservoir; 7—water reservoir; 8—air compressor.

or electricity) via energy recuperation and regeneration. We consider here two schemes of the "hybrid model of energy consumption," following the energy transfer to the shafts of the driving carts: a parallel scheme and a serial scheme.

3.3.1 Parallel Hybrid Scheme

At least two different engines are used here: thermal and electric. They are mechanically coupled, and the drive is due to the sum of their output tractions. The characteristic feature of that scheme is that even at the start of machine operation, the energy flux is divided into two autonomous parallel fluxes (Figure 3.20)

- The first one is directed through the thermal engine 1 to the shaft of the traction carts 5.
- The second energy flux is realized by a three-phase electricity generator 2 which is mechanically coupled with the thermal engine. Generator capacity is 10%–15%. The generation of electrical energy increases the efficiency of the thermal engine. Electricity is accumulated by the accumulator block 3 and then is used to drive the traction electric motor 4.

The thermal engine in this scheme is a basic traction motor whose power is found via an optimization procedure presented in what follows. (An engine operating after the diesel cycle is used as a thermal engine. Its efficiency amounts to 26%–28%). The additional electric motor is switched on when the mechanical system needs additional power (for instance, at train departure or ascent-see Table 3.2, column 5).

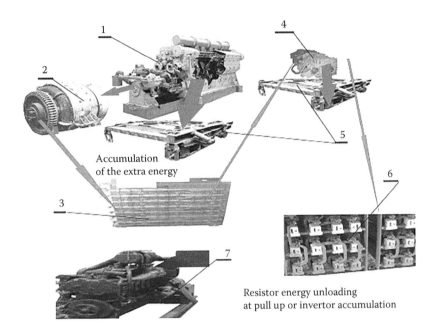

Accumulation
of the extra energy

Resistor energy unloading
at pull up or invertor accumulation

FIGURE 3.20
Arrangement of the energy fluxes in parallel driven locomotives: 1—basic traction motor (BTM); 2—electric generator; 3—accumulator block; 4—additional traction motor (ATM); 5—traction carts for the basic and additional motors; 6—resistor for dynamic pull up; 7—cart with the BTM (side view).

The diagram of the operational regimes of the equipment (including traction engines, generator, and accumulation system) during locomotive motion along a railway is also illustrated in Table 2.2. The table shows that the thermal engine operates continuously and under quasistatic loading. Assume that the engine has been correctly selected taking into account the specific loading applied.

Then, it can be exploited with minimal consumption of primary energy carriers and maximal efficiency. The table also shows that the system "generator–accumulator" (2–3) is designed to accumulate the extra mechanical energy generated by the thermal engine under all regimes, except for "departure."

The additional electric motor in some locomotives not only "amplifies" traction of the thermal engine at loading "peaks", but it can also operate jointly with the accumulator block 3 instead of using the resistor group 6 at dynamic pull up (columns 6 and 4 in Table 3.3).

The generalized energy model of a parallel hybrid system is shown in Figure 3.21. A model of classical mono-engine drives is also shown for comparison.

The following equation of energy balance is used for parallel hybrids:

$$E_M = E_T + E_E = \frac{E_{TE}}{\eta_M \eta_{TE}} + \frac{E_{EM}}{\eta_M \eta_{EM}} = \frac{\eta_{EM} E_{TE} + \eta_{TE} E_{EM}}{\eta_M \eta_{TE} \eta_{EM}}, \qquad (3.2)$$

where E_M is the amount of primary energy, necessary for the performance of mechanical work, E_{TE} and η_{TE} are the consumed primary energy and the efficiency of the thermal engine, respectively, while E_{EM} and η_{EM} are similar characteristics of the electric motor.

TABLE 3.3

Diagram of the Operational Regimes of the Parallel Drive

Motion	Basic Traction Motor TM[a]	Generator	Accumulation	Additional Motor AM[b]	Regime of Generation
On flat terrain	+	+	↑	×	×
Ascending a hill	+	+	↓	+	×
Descending	+	+	↑↑	×	+
Departure	+	−	×	+	×
Intermediate pull up	+	+	↑↑	×	+

Note: +, operates; ×, does not operate; ↑, accumulated energy; ↓, consumed accumulated energy; ↑↑, energy accumulated along 2 channels.

[a] TE—thermal engine (typical diesel engine).

[b] EM—2 ÷ 4 asynchronous electric motors with wound rotor (frequency control), adapted for generator pull up and accumulation of the generated electricity in accumulators.

Introduce the ratio between the nominal powers of the thermal engine and the electric motor $\beta_{T/E} = P_{TE}^n, kW/P_{EM}^n, kW$ as an argument of the parametric optimization to be further discussed. Then, the mechanical energy of Equation 3.2 will read

$$E_M = P_{TE} \frac{\eta_{EM} * \tau_{Op} + \dfrac{\eta_{TE} * \tau_{Inc}}{\beta_{T/E}}}{\eta_M \eta_{TE} \eta_{EM}}, \tag{3.3}$$

where τ_{Op}, h/year—operational time of the thermal engine and τ_{Inc}, h/year—operational time of the electric motor. The expression derived will be used in our economic analysis.

The plots of the energy model of parallel hybrid drives in Figure 3.21 show that the expected energy efficiency amounts to 12%–15%, while during railway maneuver it can

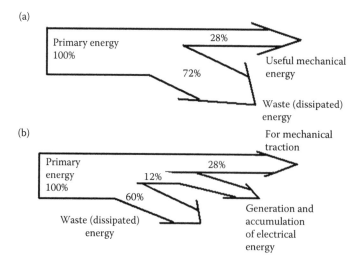

FIGURE 3.21

Energy model of parallel hybrid drive of a locomotive. (a) Classical drivers. (b) Scheme parallel.

even reach 25%–30%. The expected economy of primary energy carriers suggests that hybrid locomotives will enter railway practice through the maneuvering equipment of large stations.

3.3.2 Serial Hybrid System (Serial Hybrids)

The energy distribution scheme for serial hybrid locomotives is shown in Figure 3.22. Two types of engines are also used in its physical realization—a thermal engine and an electric motor. Yet, they are not mechanically but energetically coupled.

The energy flux successively passes through the components shown in the scheme: initially through the thermal engine 1 (Figure 3.22), which is aggregated to the electric generator 2. The primary energy carrier is transformed into electrical energy within the engine-generator group. Exhaust gases with significant energy potential (about 60%) and high temperature and capacity are recuperated by a gas microturbine coupled to the electrogenerator 5. The expected conversion of the regenerated gas micro-turbo-aggregate, operating after Brayton cycle, is about 50%–60% of the energy potential of gases flowing out of the thermal engine.

The energy model of that hybrid system is shown in Figure 3.23. It is generally assumed that even without regeneration of the combustion products, energy economy due to the flexible control of the traction energy fluxes, only, will be equal to 18%–20% for railway locomotives. If an appropriate gas microturbine is present, the economy of primary energy carriers, compared to that of classical monovalence locomotives, will reach 40%–45%, and the total efficiency of hybrids will rise from 40% to 65%–70%.

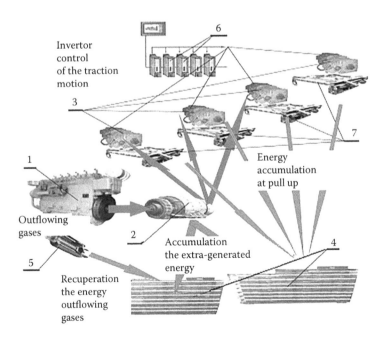

FIGURE 3.22
Energy fluxes of locomotive serial hybrid drive: 1—on-board thermal engine; 2—on-board electric generator; 3—traction electric motors: 4—accumulator blocks; 5—gas micro-turbine; 6—on-board controller for inverter control of traction motors; 7—traction carts.

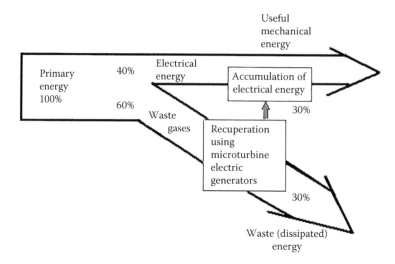

FIGURE 3.23
Energy model of serial hybrid drive of locomotives.

The generated electricity is used for the supply of traction asynchronous motors 3 mounted on carts 7. Depending on the power needed, 2, 3, or 4 traction three-phase asynchronous motors with an autonomous frequency control are used, employing the control strategy of efficiency optimization. The diagram of the operational regime of the energy equipment (including traction motors, generator, and accumulation system) is also illustrated in Table 3.4.

As seen, the thermal engine and the aggregated electric generator operate continuously. Single traction electric motors are switched on only when mechanical energy is needed (all motors are switched on under extreme loading, only, and only half of them operate under nominal motion regime).

Traction motors switch over to generation regime at the time of descent (they operate as electrodynamic brakes but electric accumulators accumulate the generated energy).

Practically, locomotives with serial hybrid drives present a rather powerful electric power station on wheels (the power of the "evolution" locomotive, for instance, is

TABLE 3.4

Diagram of the Operational Regimes of the Serial Drive

Motion	Main Engine TE[a]	Generator	Traction Motors and EM[b]	Generation Regime of EM	Accumulation
On plane terrain	+	+	++	×	↑↑
Ascending a hill	+	+	++++	×	↓
Descending	+	+	×	++++	↑↑ ↑↑
Departure	+	+	+++	×	↓
Intermediate pull up	+	+	×	++++	↑↑ ↑↑

Note: +, operates; x, does not operate; ↑, accumulated energy; ↑, consumption of the accumulated energy; ↑↑, accumulation along 2 channels.

[a] TE—thermal engine (typical diesel engine).

[b] EM—2 ÷ 4 asynchronous electric motors with wound rotor (frequency control), adapted for generator pull up and accumulation of the generated electricity in accumulators.

2387 kW*), moving with very high speed. The following balance energy equation holds for serial hybrids:

$$E_G = \eta_{TE} E_{TE} = E_{EM} + E_{AC} = \frac{E_M}{\eta_M \eta_{EM}} + E_{AC},\qquad(3.4)$$

where E_G and E_{AC} are the generated and accumulated energy, respectively.

Using again the ratio $\beta_{T/E}$, Equation 3.4 can be rewritten as

$$P_{TE}\left(\eta_{TE}\tau_{Op} - \frac{\chi_{Op}*\tau_{Op}}{\beta_{T/E}}\right) = E_{AC}.\qquad(3.5)$$

Considering that thermal engine efficiency η_{TD} is the ratio between the useful and the consumed energy, and it is identical to the so-called "real" efficiency set forth by

$$\eta_E = \frac{3.6\times10^3}{b_0 Q^P_{Down}} = \frac{3.6\times10^3 * P^{Eff}_{TE}}{m_F * Q^P_{Down}} = \frac{3.6\times10^3 * \eta_M * P^{In}_{TE}}{m_F * Q^P_{Down}},$$

consumption of the primary energy carrier per hour (mass rate), needed for the performance of the economic analysis of a specific hybrid scheme, will take the form:

$$m_F = \frac{3.6\times10^3 * \eta_M * P^{In}_{TE}}{\eta_{TE} * Q^P_{Down}},\ kg/h,\qquad(3.6)$$

where $\eta_E = \eta_{TE}$ is the real efficiency of the thermal engine; $P^{In}_{TE} = p_{In} * n_{St} * 2zV_h/N_t$ is indicative power; p_{In} is indicative pressure; n_{St} is stationary revolutions; $2zV_h/N_t$ is a drive complex comprising z cylinders, the working volume of the thermal engine V_h and the tact number N_t; Q^P_{Down}, kJ/kg is the lower heat generation capacity of the primary energy carrier (its typical value for liquid fuel is assumed to be $Q^P_{Down} = 40$ MJ/kg). We shall use these relations to estimate the economically profitable relation $(\beta_{T/E})_{Opt}$.

Serial hybrid schemes have the following advantages:

1. The thermal engine operates under quasistatic and stable conditions, where maximal conversion coefficient of the electric generator is attained (a two-cycle engine operating within the diesel engine is preferred, typical for 900 tr/min). This generation group may be replaced in the future by a high-temperature FC.

2. Traction asynchronous motors can be included in schemes for invertor control with respect to power and in those for energy recuperation at pull up.

3. Consider the parameters and capacity of the exhausting gases and note their enormous "hidden" resource in increasing locomotive efficiency. Incorporation of *gas microturbines for generation of electricity* from combustion products and subsequent electricity accumulation can significantly optimize the use of primary energy resources.

---

* General Electrical Transportation—series Evolution, maximal road speed—177 km/h.

3.3.3 Economical Aspects of the Hybrid Drives (Optimization of the Ratio between the Power of the Thermal Engine and That of the Electric Motor)

Today, transport operators are market subjects investing in transport comfort but accounting for to environment-friendliness too. They expect fast reimbursement of their investments and good dividends. Besides the ecological benefits due to the significant reduction of carbon and nitrogen emissions, hybrid technology for railway drives may have a direct economic impact reducing expenses for primary energy sources. The economy could measure up to 17%–20% for railway construction and up to 30%–35% for railway shunting.

The basic problem of the choice of a hybrid locomotive is the value of the ratio between the thermal and the electric power $\beta_{T/E}$. The rational approach for the calculation of this quantity requires the use of economic methods and criteria. We propose here a method of finding the optimal power ratio $\beta_{T/E}$, based on the dynamic method of assessing investment cash flows known as "net present value—NV_C." This method of optimization of the power ratio is based on an economical analysis. Its task is to find the sufficient hybrid (coupled) power needing minimal initial investments and allowing minimal expenses for primary energy.

A procedure observing the condition of minimal work for locomotive manufacture and operation is applied. Thus, a condition for the *minimization of the discount "NV_C"* is set forth

$$NV_C = \sum_{i=TD,ED,G,AB}^{4} K_i + \sum_{j=1}^{4} T_j e_j^{-1} \to \min, \qquad (3.7)$$

where

- $\sum_{i=1}^{4} K_i$, \$(Eu) are the single expenses K_{TD}, K_{ED}, K_G and K_{AB} for the delivery of thermal engine, electric motor, generator, accumulators, and operating electronics, respectively. Traditionally, values K_i are determined by the capacity of the respective equipment. They can be expressed by the following generalized relation: $K_i = C_i * P_i$, $i = TD$, ED, G, AB. Coefficients C_{TD}, \$(Eu)/W; C_{ED}, \$(Eu)/W; C_G, \$(Eu)/W, and $C_{AB} = kC_{AB}^C$, \$(Eu)/kW (k, Ah/kW—transfer factor) can be found performing a marketing study of equipment costs or offers.

- $\sum_{j=2}^{3} T_j e_j^{-1}$, \$(Eu) are expenses for current repairs, delivery of energy carriers, and staff salaries discounted at the instant of initial investment (amortization expenses for current repair T_1 are not subject to discount since pursuant to the national accounting standards, they do not form a real cash flow).

- e_j^{-1} is a discount coefficient of the types of future repairs expenses as compared to the initial cost ($e_j^{-1} = [1-(1+r_j)^{n_i}]$, and r_j is a risk norm, n_j is a normative life cycle of the respective system of buildings). The risk norm is found via a model of assessing the capital assets (CAPM $\to r_j = f_r(r_m - f_r)\beta$).

A practical approach to calculate the discount coefficients is to assume that the condition $r_j = i$, is satisfied, where i is the level of official inflation (inflation index).

To find the profitable power ratio $\beta_{T/E}$ we should nullify the first derivative of the functional $NV_C = NV_C(\beta_{T/E})$:

$$NV_C' = \frac{d}{d\beta_{T/E}}\left[P_T\left(C_T + \frac{1}{\beta_{T/E}} \sum_{i=E,G,AB}^{3} C_i \right) + \left(m_F * v_F * e_F^{-1} + P_T * Z_P \right) \right] = 0, \qquad (3.8)$$

where

- $\sum_{i=T,E,G,AB}^{4} K_i = P_T \left(C_T + 1/\beta_{T/E} \sum_{i=E,G,AB}^{3} C_i \right)$ are the single expenses
- $\sum_{j=3}^{4} T_j e_j^{-1} = \left(m_F * v_F * e_F^{-1} + P_T * Z_P * e_P^{-1} \right) Eu$

Here, $F_{T_2} = \left(m_F * v_F * e_F^{-1} \right)$ is the discounted value of the primary energy carrier and $F_{T_3} = P_T * Z_P * e_P^{-1}$ are the discounted expenses for salaries of the locomotive team and auxiliary staff.

The average annual consumption of primary energy carrier m_F, t/year is estimated based on the predicted amounts considering the hybrid scheme (Equations 3.2 and 3.3) and the specific terrain of the chosen route (a number of pull ups, waits and descents, other specific speed restrictions, etc.). The value of the energy carrier (v_F, \$(Eu)/t) is calculated accounting for the variable trend of the expected prices of the regional oil stock markets.

Finding the first derivative of the functional (3.8) with respect to $\beta_{T/E}$ and nullifying it, we obtain the following formula for the economically profitable power ratio:

$$(\beta_{T/E})_{Opt} = \sqrt{\frac{P_T \sum_{i=E,G,AB}^{3} C_i}{v_F * e_F^{-1} * P_T dm_F / d\beta}} = \sqrt{\frac{\sum_{i=E,G,AB}^{3} C_i}{v_F * e_F^{-1} * dm_F / d\beta}}. \tag{3.9}$$

The above formula for $(\beta_{T/E})_{Opt}$ illustrates the quadratic dependence between the nominal powers of the basic (thermal) engine and the additional (electric) motor, on one hand, and the market prices of the power equipment $\sum_{i=E,G,AB}^{3} C_i$, the discounted energy cost $v_F * e_F^{-1}$ and the specific terrain conditions implicitly present in the derivative $dm_F/d\beta$, on the other hand. Note that $dm_F/d\beta$ should be defined depending on the actual design requirements. Hence, the basic requirement to the delivery of hybrid locomotives should be the performance of a detailed economic analysis of their route of operation. This will guarantee a minimal time of investment reimbursement.

Hybrid drives used in railway, railway-maneuvering, or purely maneuvering locomotives combine the positive aspects of the thermal engine and the electric motor, namely

- They admit flexible automatic regulation via digital systems and thus attain significantly more effective use of the primary energy carrier and significant energy economy by elimination of the human driving factor (the expected economy for railroad locomotives alone is about 17%–22%).

- They guarantee steep slopes ascending (about 40%–45%) with nominal speed (up to 250 km/h) under optimization of traction and energy consumption. The new traction asynchronous motors with frequency control and efficiency of 95% save energy and provide high railway passability.

- They possess higher relative power as compared to single source drives (diesel or electric ones).

The second positive quality of hybrid drives is the high autonomy of the transport operators and their rolling stock, and the devices do not admit deterioration of the central power supplied. At the same time, their maintenance and utilization are more complicated, making respective demands to the quality of training of railway and maintenance staff.

Finally, it is seen that the adoption of hybrid locomotives by local railway transport requires higher (by 20%–25%) initial investments. Note that powerful investors or the state itself can provide such funding.

3.4 Control Questions

1. What is the difference between the direct and indirect technologies of generation of electricity?
2. Specify the basic direct technologies.
3. Can a body with wide banned zone be internally ionized? What does body internal ionization consist of?
4. What is the structure of a photovoltaic cell?
5. Specify the coefficient of conversion of the isomorphic photovoltaic cells. Specify that of polycrystalline photovoltaic cells.
6. Which is the physical phenomenon that is the basis the Seebeck effect?
7. Can one prepare "perpetuum mobile" using a single thermocouple?
8. Specify the thermocouple components.
9. How does a thermoelectric converter operate? In what area can it be applied?
10. Which are the factors affecting the conversion coefficient of the Seebeck technology?
11. What is the difference between the effects of Seebeck, Peltier, and Thompson?
12. Where are the effects of Seebeck/Peltier applied? Specify an example.
13. What is the difference between the spontaneous and controlled oxidation of energy carriers? How does the controlled oxidation proceed technologically?
14. Which are the basic types of low-temperature FCs? What fuel do they operate with? Can they work using natural gas and methane?
15. What is the maximal voltage generated by a FC? What is the electrical efficiency of low-temperature cells?
16. How could one use low-temperature cells for the energy supply of buildings?
17. What is the energy efficiency of low-temperature FCs under energy co-generation?
18. Which FCs are high-temperature ones? Which are the carriers of energy through the electrolyte of solid-oxide and MCFCs?
19. What fuels are used in high-temperature FCs? What is their electrical efficiency? What is their efficiency under the regime of co-generation?
20. Specify the advantages and the disadvantages of low-temperature FCs.
21. Give examples of the application of high-temperature FCs.
22. What are the differences between technologies employing direct and indirect energy conversions?
23. Which are the two most popular indirect technologies of electricity generation?
24. Why has the Rankine technology developed prior to the industrial development of the gas-turbine technology?

25. Which are the basic similarities between the steam-compressor and gas-compressor thermoelectric technologies?

26. Specify the basic advantages of the Brayton thermoelectric technology.

27. Why was Faraday's generator not applied industrially?

28. Why did William Grove's technology, although its almost half-century popularity, give way to steam-turbine technology?

29. How can a steam-turbine installation be adapted to operate under co-generation? What will be the rate of its efficiency increase?

30. How can an old jet engine be used as a gas-turbine system for the generation of electricity?

31. How can that engine be adapted to operate under co-generation?

32. Consider a multistage compressor employed by the Brayton thermoelectric technology. How would intermediate air cooling affect energy efficiency? Why?

33. Outline the areas of application of the Rankine technology, when heat is generated using renewable sources.

34. Outline vehicles where steam-compressor technology is applicable.

35. Give examples of co-generation in atomic power stations.

36. What thermal technology would you use in the design of a plant for "daily waste utilization"?

37. Propose a scheme of in-depth utilization of agricultural bio product waste.

38. How could you arrange a project for energy supply of a farm?

39. Where in transport could thermoelectric technologies be applied?

40. What is the principal difference between a serial and a parallel hybrid scheme of vehicle drive?

41. When does a parallel hybrid technology outrun a serial hybrid technology?

42. How can efficiency of the serial hybrid technology be additionally increased?

43. Specify a scheme of utilization of combustion products; exhaust the thermal engine of a hybrid drive.

44. What are the restrictions imposed on the operation of combustion products potential from designer's point of view?

4

Energy Rehabilitation: Regeneration and Recuperation

4.1 Energy Conversion as Interpreted by the Principles of Thermodynamics

Not long ago, when the cost of primary energy carriers was low, energy waste (low-potential heat, mostly) was neglected in the energy balances of most manufacturing technologies.[*] Yet, the market price[†] of fossil energy carriers changes, when their consumption increases presenting the gloomy possibility of less oil being used when oil is one of the most preferred energy carriers. On the other hand, changes in the global climate and ecology urged the public, investment, scientific, and engineering interest in technologies that would more effectively exploit the potential of the primary energy carriers by increasing the efficiency of the technologies of energy regeneration and recuperation.

A number of such technologies have been developed and applied in power engineering practice. Generally, they are known as "energy conversions," and they are classified in two groups as shown in Figure 4.1:

- Regeneration technologies are those where the energy potential of the working fluid is used directly or increased by keeping the energy form—generally by increasing the temperature or pressure of the authentic energy-charged medium
- Recuperation technologies are those where the available energy potential of the working fluid is used to change the energy form (for instance, "heat or mechanical motion to electricity" (see Figure 4.1) at the expense of a more complete use of the residual enthalpy[‡]

Classical thermodynamics assumes that the energy of a system should be estimated regarding the value of entropy (*s*). When entropy in an energy-converting system (ECS) is small, its energy charge is considered to be large and it is able to operate (doing mechanical, electrical, chemical, and other kinds of work). Systems with great entropy

---

* Exceptions were made in metallurgy, silicate industry, and energetics where various regeneration solutions were applied. For instance, boilers, using the combustion products for production of hot water were incorporated to the open-heath furnaces.
† Actually, the market prices of oil and natural gas depend on a set from geopolitical and geoeconomical factors.
‡ Except for enthalpy, another term for the assessment of energy potential is "the free energy potential or exergy."

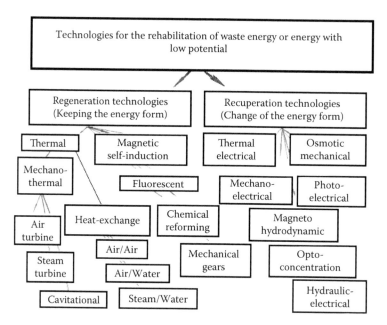

FIGURE 4.1

Classification of the technologies for energy rehabilitation: • regeneration technologies and • recuperation technologies.

are energetically "discharged" and not fit for work since they are in equilibrium. Classical thermodynamics assumes that availability of large amount of entropy proves that the system is charged with energy of "poor quality" or energy waste, and it is de-energized. Hence, the change of entropy (a thermodynamical system [TDS] state parameter) is an important indicator (key) to possible energy conversion in the equipment and aggregates involved in a certain manufacturing process.

Note that a dynamic balance of energy exists in an isolated TDS (based on the *First Principle* of Thermodynamics) and all energy conversions naturally yield entropy increase and equilibrium in the control volume (based on the *Second Principle*). The development of an energy-conversion process in a TDS can be assessed at two levels:

- At macroscopic level
- At microscopic level

Classical thermodynamics treats the problems at the macroscopic level by designing the pattern of the equilibrium states. Plots of the variation of TDS basic state parameters are used for that purpose (temperature and pressure, first of all, and the electric potential, if necessary). Using those one can assess the final results of the thermodynamic conversions that accompany independent thermodynamic processes or cycles. Clausius formulas for entropy are used to express the change of energy potential:

- $ds = \delta q/T$—for very small increments
- $\Delta s = \Delta q/T$—for finite variations

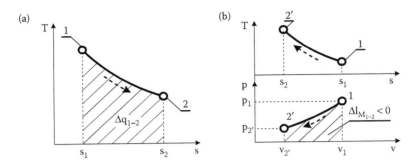

FIGURE 4.2
Entropy growth: (a) spontaneous change of entropy (always growing) and (b) controlled change of entropy (by the "introduction" of mechanical work).

When a TDS changes its state from position 1 to position 2 (see Figure 4.2a), the final value of entropy s_2 is found by integration of the expression

$$s_2 = s_1 + \sum_{1}^{2} \frac{\Delta q}{T},$$

where s_1 is the entropy initial value while the sum is an indicative characteristic of the system de-energization. Two approaches are used to find entropy s_2—analytical and semiempirical.

The first one adopts the differential form of heat variation δq to be expressed by the change of the internal energy (du) and the introduced mechanical work (δI_M) applying the law of energy conservation $(\delta q = du + \delta I_M = C_v dT + p dv)$.

Next, the thermodynamic restrictions following from the equation of state* or from some of its special applications are imposed, and they are presented by a polytropic law $(pv^n = RT)$.

The second approach requires current measurement of temperature T and the amount of released or supplied heat (Δq). Those quantities should be experimentally found while entropy should be calculated.

Applying the first approach, we find that the entropy final value s_2 is

$$s_2 = s_1 + C_v \ln \frac{T_2}{T_1} + R \left(\frac{v_2}{v_1} \right)^{1-n}$$

Hence, we may predict that entropy s will grow if the following condition is fulfilled:

$$\ln \frac{T_2}{T_1} + \frac{R}{C_v} \left(\frac{v_2}{v_1} \right)^{1-n} > 0$$

---

* The Clapeyron equation is valid for equilibrium states.

or if the following inequalities hold true:

$$T_2 > T_1 \quad \text{and} \quad v_2 > v_1 \tag{4.1}$$

or

$$T_2 < T_1 \quad \text{and} \quad \frac{R}{C_v}\left(\frac{v_2}{v_1}\right)^{1-n} > -\ln\frac{T_2}{T_1}. \tag{4.2}$$

Those conditions can be physically fulfilled if a TDS is supplied with heat, and the TDS performs mechanical work. Entropy increases with an increase of the TDS control volume ($dv > 0$) or decrease of temperature ($dT < 0$), but inequality (4.2) should be always satisfied under natural conditions.

It follows then in a TDS expansion (i.e., performance of mechanical work) entropy will always increase until the equilibrium is reached. The TDS will have greater entropy at the new equilibrium state determined by the new volume v_2, and the potential of the TDS energy charge will be lower since the system has lost its working capacity (the term "*de-energized TDS*" is regularly used). Yet, the details of the system "energy reduction" are not clear from a physical point of view. To illustrate that process, we use the tools of a new thermodynamic science called "statistical thermodynamics." It emerged at the end of the 19th century and the beginning of the 20th century owing to the works of L. Boltzmann (1844–1906), J. Maxwell (1831–1879), A. Einstein (1879–1955), S. Bose (1894–1974), E. Fermi (1901–1954), and a number of other researchers.

The basic approach of this trend is the analysis of the processes at micro-level, assessing the behavior of the individual energy carriers during their interaction with each other via statistical data processing. The approach main advantage is that the general picture (the macroscopic state) is drawn by considering all possible microstates, considering the history of energy conversion.

Microstates are transitory dispositions of system-compounding particles that "carry" the system energy (photons, phonons, electrons, and ions). When a system is in equilibrium, all TDS microstates are similar to each other, and are statistically indiscernible. Transition from one microstate to another takes place "automatically" and within one and the same time intervals ($\sim 10^{-15}$ s) or under constant frequency ($v = \tilde{\Gamma}^{-1}$, Hz), where $\tilde{\Gamma}$ is the number of microstates. When a TDS undergoes energy conversion after a supply of external work, the single microstates become statistically discernable.

The entire set $\tilde{\Gamma}$ of TDS microscopic states a_i is called "microscopic canonical ensemble." The system automatically runs over all microscopic states. Inspecting the behavior of a TDS, one can find that a group of microscopic states exists, which is characterized by particle identical behavior and characteristics. This group forms the so-called "collective" macroscopic state (or macrostate *No* 1). Similarly, other macrostates also exist but as compared to macrostate No 1, they contain much less microstates. They are called minoritarian macrostates No 1, No 2, etc. The evolution of the energy conversion is followed by as evaluating the relation between the minoritarian macrostate and the total macrostate.

When the number of minoritarian macrostates increases, the time of TDS "occupation" of each of them increases. Note that the frequency of TDS occupation of a certain minoritarian macrostate also increases. At the macroscopic level, this will cause a change of the values of the measured parameters, and the registered values of temperature (T), pressure (p) and electric potential (V) will also change.

To get an idea on the "de-energization" of an insulated TDS,[*] assume that its state at point 1 (Figure 4.2a) is defined by a specific volume v_1 (small value), entropy s_1 (small value), pressure p (high value), and temperature T (high value), that is, the TDS is energetically charged. In statistical thermodynamic, a TDS state is specified by the form of the energy-distribution function (the energy spectrum). Regarding the case under consideration, the energy spectrum at point 1 is characterized by domination of a specific energy form which is common for all energy carriers:

- If the working medium is a solid body, the energy-carrying photons have identical frequency and transport direction
- If the working medium is an ideal gas, all its molecules move with identical velocity (regarding magnitude and direction) and possess identical kinetic energy, respectively

The system thus described (characterized by the parameters at point 1) is in equilibrium, and its individual microstates are indiscernible between each other and with respect to the collective macrostate. Change of the state of an insulated TDS can occur spontaneously, after an instant expansion of its control volume (v),[†] for instance.

The statistics of the development of new microstates admits the introduction of *another interpretation* of entropy owing to L. Boltzmann (1874–1896)—a German physicist, natural philosopher, and founder of statistical thermodynamics. He states that *entropy* s_i is proportional to the thermodynamic probability of the existence of a certain microstate within the "microscopic canonical ensemble" written in the following form:

$$s_i = \kappa_B \ln(\tilde{\Gamma}_i)\, J/K, \qquad (4.3)$$

and $\tilde{\Gamma}_i$ is the number of current microstates in the *i*th minoritarian macrostate while $k_B = 1.38 \times 10^{-25}$ J/K is the Boltzmann constant.

Entropy of the entire system S_{sys} is the normalized mean sum of entropies of all n_M minoritarian macrostates within the canonical ensemble, that is,

$$S_{sys} = \frac{1}{n_M} \sum_{i-1}^{n_M} s_i. \qquad (4.4)$$

The analysis of expression (4.3) shows that the system entropy increases when new forms of motion of energy carriers and new minoritarian TDS macrostates occur, whereas the number of factor s_i in Equation 4.4 increases.

---

[*] In adiabatic conditions, $\delta q = 0$ when the working fluid moves in an intake nozzle, between the blades of gas or steam turbines or through throttle-valve, etc.

[†] When the control volume expands (dv > 0) the kinetic energy of part of the energy carriers decreases. 3D effects occur (including turbulence). This widens the spectrum of energy forms and other microstates emerge. Their share progressively increases yielding a new collective macrostate and new equilibrium of the system.

This is observed in all the cases of adiabatic, isothermal, and isobaric expansions which are characteristic for insulated TDS, and it is explained by the emergence of new *3D* energy forms at microscopic level. Mechanical work for TDS expansion is done at macroscopic level. Hence, TDS should be de-energized and the energy potential should decrease.* Formally, this is also accompanied by increase of entropy S_{sys} (basically, by decrease of the temperature[†] of the ideal gas).

Nothing of the sort happens when a TDS *"contacts"* a hot energy source thus receiving external heat, although that entropy increases. The TDS may expand (and do mechanical work) or its total internal energy (enthalpy) may increase. This, however, does not deteriorate the energy potential but on the contrary-increases the TDS working capacity although new energy forms emerge meanwhile, generated by the TDS interaction with the hot body. A classical example in this respect is the entropy increase during isochoric heating (for instance, a process running along the line "B-1" in Figure 2.2) where heat q_1 is introduced in the TDS without doing mechanical work (the TDS boundaries do not move). Yet, new microstates occur related to the emergence of energy carriers, and their internal energy is higher than that of the states of the "collective" microstate.

Here, the emergence of a new form of motion is due to the "explosive" running of the chemical reactions within the total volume of the combustion chamber of the carburetter engine. Yet, in other cases, this may result from the performance of thermal work via transfer of long-wave phonons along the TDS boundaries (as is in the Stirling engine). Gradually, those phonons approach the kernel of the TDS control volume and hence, the system working ability grows as a whole. Parallel to the "shift" of the energy carriers to new energy orbitals, the number of minoritarian TDS states increases, and thus entropy also increases (see Equation 4.4). Although that the system attains equilibrium at larger entropy S_{sys} (at the limit point 2 in Figure 2.2), its energy charge is significantly larger since temperature (T_2) and internal energy (u_2) have increased.

In most real cases of energy supply of an insulated TDS, the volume of the working fluid increases ($dv > 0$ and $l_{M_{B-1}} > 0$). Although that entropy grows ($S_{sys}\uparrow$), the energy charge of the system also grows. The same effect is observed when the TDS shrinks its volume (see Figure 4.2b) under mechanical external work ($l_{M_{B-1}} < 0$). Consider for instance isobaric heating (p = const). Then, heat supplied is spent to increase the TDS total energy (enthalpy). Similar is the behavior of a TDS under polytropic heating.

Those observations show that to objectively estimate the quality and working capacity of waste energy in industry, daily life, and transport, one should not count on entropy as an unique indicator. From a practical point of view, one should be interested in TDS "deprival of energy," since the technologies of energy regeneration and working capacity restoration can then be applied. Hence, it seems useful to introduce a *specific measure* of TDS degree of de-energization, defined by the ratio between the exergy decrease in a certain technological process or device $\Delta D = (D_1 - D_2)$ (see Table 4.1) and the exergy initial value

$$IEE_{EC} = \frac{\Delta D}{D_1} = 1 - \frac{D_2}{D_1}. \tag{4.5}$$

---

* An alternative assessment may be performed using the exergy $B = u + p.v - T.s + V.e$ which contains information on the total thermodynamic cycle.

† In real working bodies for $T < T_{inv}$, see Section 1.5—the effect of Thompson.

TABLE 4.1

List of the Regular De-Energizing Processes

Process Measure	Entropy	Temperature	Decrease of Exergy $\Delta D = \Delta h - vdp$ $\Delta D = \Delta u - pdv$
Adiabatic expansion or cooling	$s_1 = s_2 = \text{const}$	$T_2 = T_1 - \dfrac{l_{M_{1-2}}}{C_v} \downarrow$	$D_2 = D_1 - l_{M_{1-2}} \downarrow$
Isochoric cooling	$s_2 = s_1 - C_v \ln(T_1/T_2) \downarrow$	$T_2 = T_1 - \dfrac{q_{1-2}}{C_v} \downarrow$	$D_2 = D_1 - C_v(T_1 - T_2) \downarrow$
Isothermal expansion $(q_{1-2} = l_{M_{1-2}})$	$s_2 = s_1 - \dfrac{l_{M_{1-2}}}{T_2} \uparrow$	$T_1 = T_2 = \text{const} = T$	$D_2 = D_1 - l_{M_{1-2}} \downarrow$
Isobaric expansion or cooling	$s_2 = s_1 - C_p \ln \dfrac{T_1}{T_2} \downarrow$	$T_2 = T_1 - \dfrac{q_{1-2}}{C_p} \downarrow$	$D_2 = D_1 - C_p(T_1 - T_2) \downarrow$
Polytropic expansion or cooling	$s_2 = s_1 - \dfrac{n-k}{n-1} \ln \dfrac{v_1}{v_2} \downarrow$	$T_2 = T_1 - \dfrac{n-k}{n-1} \dfrac{q_{1-2}}{C_v} \downarrow$	$D_2 = D_1 - (u_2 - u_1) - l_{M_{1-2}} \downarrow$

It is called index of efficiency of the energy conversion (or index of regeneration/recuperation processes). This quantity is suitable to assess both specific regeneration technologies and instruments and installations for technology operation and labeling.

A higher IEE_{EC} yields more efficient use of the energy potential of the system and higher degree of system de-energization. The maximal limit value of $IEE_{EC}(IEE_{EC} = 1)$ is attained when $D_2 \Rightarrow 0$ or when the potential of the free energy is removed.

The energy of the working fluid assessed by means of the exergy at the exit of the installation or energy device D_2 is the so-called "energy waste," "low quality" energy, or "poor" heat, being shot into the surroundings. The energy carriers of a TDS in such a state occupy low-energy orbitals and they are not "work-capable."

Table 4.1 specifies a list of those processes, and the TDS becomes de-energized after their conclusion, thus releasing energy waste. The forms of the exergy, assessed at the exit of the installation or energy device D_2 and specified in column 4, show that the exergy absolute value decreases for all processes in Table 4.1. Hence, the following conclusions concerning residual energy quality and possibilities of its transformation should be made:

1. The residual energy in a noninsulated TDS can not be qualitatively assessed using the TDS entropy S_{sys}. This follows from the second principle of thermodynamics. Yet, the assessment can be made using the TDS *free gross potential* known as exergy and following exergy development in time.

2. The restoration of the working capacity of the residual energy can only be done by the introduction of *external work* into the TDS. This can be mechanical, chemical, electrochemical or thermal work, or a combination of them. In what follows, we shall describe and discuss some recuperation and regeneration technologies with various applications in industry and constructions, following the order in Figure 4.1.

4.2 Regeneration Technologies

Generally speaking, regeneration of waste energy is restoration of its working capacity while keeping its form (see Figure 4.1). More specifically, heat regeneration is a technology of heating water, which flows into a boiler (steam generator) or fresh air, which is supplied to gas turbines, air conditioning, or ventilation systems by waste heat from combustion products or utilized air, respectively.

Those processes are plotted as a so-called regeneration cycle (see in Figure 4.3a), consisting of two isothermal and two polytropic processes ($n_1 = n_2 = $ const). Temperatures T_1 and T_2 are very close to each other and energy exchange takes place under a very small gradient. Both processes can run simultaneously (i.e., synchronously) or they can be separated in time.

Energy absorption and release takes place in a device called regenerator. Its various designs are shown in Figures 4.4 and 4.5. Regardless of the fact that the processes of energy exchange are non steady, similar processes run in both regenerators:

1. The heat carrier with temperature T_2, charged with energy waste, "cools down" along the polytrope 2-B to temperature T_B, releasing heat q_R to the heated carrier

2. The heated carrier with temperature T_1 absorbs heat along the polytrope 1-A and heats up to temperature T_A

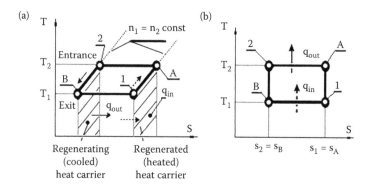

FIGURE 4.3
Counter clockwise regeneration cycles: (a) regeneration without the introduction of external mechanical work and (b) regeneration with the introduction of external mechanical work (counter clockwise Carnot cycle).

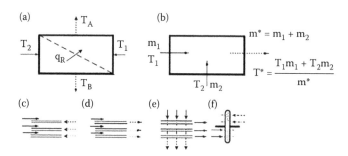

FIGURE 4.4
Steady convective heat exchangers: (a) with solid walls; (b) mixing; (c) counter flux; (d) collinear flux; (e) crossed flux; and (f) heat pipes.

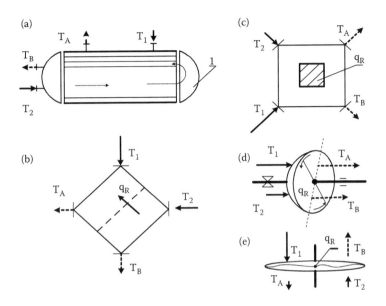

FIGURE 4.5
Structure of regenerating heat exchangers: (a) with casing and pipes; (b) with plates; (c) accumulation (latent and entropy) one; (d) hot wheel; and (e) with a heat pipe.

As a result of the heat exchange, part of the energy waste (q_R) of the cooled carrier is transferred to the heated carrier. It enters the regenerator with temperature T_2 and leaves it with temperature T_B ($T_B \approx T_1$).

Using relation (4.5), we can find the efficiency of the energy waste regeneration

$$IEE_{EC} = 1 - \frac{D_B}{D_2} = u_2 - u_B + l_{MB-2}$$

$$= C_v(T_2 - T_B) + \frac{R}{n+1}(T_2 - T_B)$$

$$= \left(C_v + \frac{R}{n+1}\right)(T_2 - T_B),$$

since $T_B = T_1$ (here n is the polytrope exponent).

As the analysis of the expression for IEE_{EC} shows, regeneration will be more efficient when the difference ($T_2 - T_B$) is high, which is achieved by deep cooling of the energy carrier (to the lowest possible temperature T_B).

The characteristic feature of the regeneration cycle thus described is that it is realized under small temperature difference ($T_1 - T_2$). To be more effective in real conditions, the rate of heat exchange between the energy "donor" at lower temperature and the energy "recipient" (the TDS working fluid)* this difference should be larger.

---

* In this respect, regeneration carried out by means of heat pipes is the most effective one (the rate of heat transfer may reach 800–1000 W/m²K).

4.2.1 Heat Exchanging Devices

4.2.1.1 Heat Regenerators

The general mechanisms of heat transfer were discussed in the previous paragraphs, while the treatment of the control volume and the TDS boundaries was abstract or simplified. Note that heat exchange in energy equipment is carried out in special devices called heat exchangers. Heat exchange can take place there with or without mass exchange (without or with mixing of working fluids with different temperatures).

Accounting for that feature, heat exchangers can be divided into regenerative (Figures 4.4a and 4.5) and mixing types (Figure 4.4b). Energy release and absorption in these two devices take place under steady conditions, simultaneously in one and with the same physical volume.

Regenerating heat exchangers perform exchange of heat between working fluids with different chemical composition. These are for instance combustion products and water, refrigerating agents, air, superheated steam etc., used in industry or daily life. The effective area of the heat exchanger is assessed via the expression

$$A_{eff} = \frac{Q}{U \cdot \Delta T_m}, \, m^2, \tag{4.6}$$

where

- Q, W—flux of the available low-potential heat suitable for regeneration
- U, W/m²K—transitivity coefficient or coefficient of heat transfer (see Table 4.2)
- ΔT_m—logarithmic mean temperature difference

Performing transformation of the above expression, we get a formula to assess the temperature of the working fluid after regeneration accomplishment:

$$T_A = T_B + \frac{Q}{(1/(U * A_{eff}) + 1/2\overline{W}_1 + 1/2\overline{W}_2)},$$

where

- $\overline{W}_i = C_p m_i^*$—i = 1.2 is water equivalent of the mass rate m_i^*, kg/s
- T_A and T_B are exit temperatures of the regenerated and regenerating working fluids

Other regenerators, yet operating under unsteady conditions, are those shown in Figure 4.6. The first two in Figure 4.6 have a rotating hot wheel 1 which receives and accumulates energy via heating during its passage through the hot (regenerating) fluid 2 (combustion products, cooled air, wet steam, etc.). When the already hot wheel 1 starts

TABLE 4.2

U-Value Coefficients for Water to Water or Water to Steam Heat Exchangers

| Type | Water/Water | | Steam/Water | |
	Speedy	Volumetric	Speedy	Volumetric
U, W/m²K	1000–2000	250–400	2000–3500	400–1200

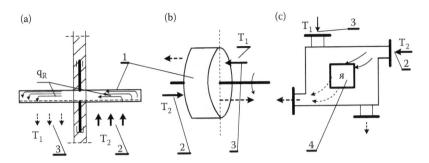

FIGURE 4.6
Unsteady convective heat exchangers: (1) hot wheel; (2) HS; (3) CS; and (4) accumulating core. (a) Vertical rotating wheel; (b) horizontal rotating wheel; and (c) heat trap.

rotating and enters the area of the cold (regenerated) fluid 3, it releases the accumulated heat to the fluid due to the temperature gradient.

The operation of regenerator in Figure 4.6c is unsteady. Yet, phases of energy accumulation and consumptions are shifted in time (for instance, cheap energy is accumulated at night and expensive energy is consumed during the day). Note the necessity of introducing a unified system for the assessment of energy efficiency of the regeneration equipment as a conclusion. A leading example in this respect is the certification (labeling) system for household equipment.

In some cases of regeneration of low-potential heat, energy efficiency can be significantly increased by increasing the temperature T_2 of the working fluid. From a theoretical point of view, this can be achieved by a specific arrangement of the regeneration cycle, including the introduction of additional energy into the heat exchanger in the form of mechanical or chemical work (by compression or combustion and absorption,* for example). In fact, the idea of regenerating energy waste from industrial technologies by doing mechanical work is rather old. The work done is used to increase the enthalpy of the working fluid. This is proved on macroscopic level by the increase of body pressure, density (decrease of the specific volume, respectively), and temperature.

On a microscopic level, the direct result of that work is proved by (i) increase of the internal potential energy owing to the decrease of the distance between the atomic nuclei and (ii) the increased number of collisions between the energy carriers due to the decrease of the physical volume of the TDS control area.

Practically, the "mechanical regeneration" is performed by compression of the working fluid along a counterclockwise Carnot cycle, whose diagram is shown in Figure 4.3b.

As known from the theory of the counterclockwise cycles, the mechanical work done over a TDS for its compression is larger than the work needed for TDS expansion, and the result is that heat (q_1) released within the hot body exceeds heat (q_2) received by the TDS during a contact with the cold body.

Regeneration devices called heat pumps[†] are designed to operate within a cycle consisting of four processes: two isothermal processes (for T_1 = const and T_2 = const) and two adia-

---

* The devices, using chemical work for energy regeneration should be not referred here.
† The term "heat pump" is introduced owing to the direction of motion of the energy flux. That device "receives" heat (q_2) from the cold source. Then, external mechanical work (L_C) is "inserted" by compression of the working fluid and the system temperature increases to T_2. Energy (q_1) is released into the surroundings.

batic processes (for s_1 = const and s_2 = const). They are identical to those of the Carnot ideal heat engine. In contrast to it, however, the heat pump cycle operates counterclockwise starting from point 1 (Figure 4.3b).

Efficiency of the counterclockwise Carnot cycle of the heat pump is usually assessed via the coefficient of performance specified as

$$\beta_{HP}^{Theoretical} = \frac{\text{Useful effect}}{\text{Inserted external work}} = \frac{q_{out}}{L_u} = \frac{1}{1 - T_1/T_2} > 1.$$

The coefficient of performance β_{mn} is larger than 1 ($T_1/T_2 < 1$), and it proves the advantages of heat pumps in buildings and industry, where free energy q_2 is utilized. Only the *energy necessary for compression* from point 1 to point A^* should be paid for. The necessary work is done by an electric motor or heat engine that drives the heat pump. Theoretically, the coefficient of performance $\beta_{HP}^{Theoretical}$ depends on the ratio between the temperature of the cold source (CS) (T_1) and that of the hot source (HS) (T_2). It increases with an increase in temperature T_1 and decreases with decrease in temperature T_2.[†]

Manufacture of Carnot heat pumps and attaining a coefficient of performance within a specific temperature range is an engineering impossibility which has not been physically and commercially realized. The calculated coefficient $\beta_{HP}^{Theoretical}$, however, is used as ultimate standard of energy regeneration. It indicates the difference between the ideal concept and the coefficient of performance of real industrial devices after the inverse thermodynamic cycle.

Nowadays, two *real heat pump technologies* are popular on the market and used to perform "mechanical" regeneration of low-potential (waste) heat. They are

- Air-compressor technology
- Vapor-compressor technology

The first one employs the counter clockwise Brayton cycle while the second one—the counter clockwise Rankine cycle. Both counterclockwise cycles consist of two adiabatic and two isobaric processes. Their configuration is similar to that of the classical regeneration cycle—see in Figure 4.3b. The formal physical difference between them is based on the working fluid used. As in direct thermodynamic cycles, the counterclockwise Brayton technology uses gas (air). The counterclockwise Rankine technology uses a working fluid that experiences a phase change (refrigeration agents: chlorofluorocarbons, ammonia solutions, bromine–lithium solutions, etc.) which undergo phase transformations.[‡]

In what follows we shall consider each of the heat pump technologies used for the regeneration of waste heat or low-potential energy from renewable sources.

---

[*] No comment on the value of the initial investment.
[†] Although that η_{HP}^{Ther} may attain values of up to 13–14, the real conversion coefficients are within the range $3.5 \leq \eta_{HP} \leq 4.5$.
[‡] Heat pumps operating in the counter clockwise Rankine cycle use smaller flows of the working fluid (m³/s), since the "transport" capacity of the latter increases during phase transformation.

4.2.2 Air-Compression Heat Pump, Operating of the Brayton Counter Clockwise Cycle

Waste heat is released by various sources—geothermal areas, hot air areas, industrial technologies, or biotechnologies. To restore its working capacity, one needs to increase the temperature of the working fluid. A heat pump aggregate is used for that purpose.

It consists of air turbo-compressor (C) combined with a motor (M), air turbine (T), and heat exchangers (type "air to air" and "air to water"—see Figure 4.7a). The heat pump counterclockwise thermodynamic cycle is illustrated by the mechanical p-v and thermal T-s state diagrams of the working fluid–air (in Figure 4.7b and c).

The cycle starts with adiabatic compression (from point 1 to point A) taking place in the turbo-compressor (C), where the mechanical work is converted into internal energy and the temperature of the working fluid increases from T_1 to T_A. Then, isobaric cooling in the heat exchangers of building installations (heating installations (HI) or air conditioners (AC)) follows where the regenerated heat is released, and next comes the adiabatic cooling/expansion in the air turbine (T). The working fluid is cooled down to T_B and with decreased pressure p_B is discharged into the cold energy source (the atmosphere) to complete the cycle.

The regeneration of the "waste" energy q_2 needs the introduction of external mechanical work. Its amount per TDS unit mass is specified by the total work needed for its compression from point 1 to point 2, minus the work done within the air turbine

$$L_{In} = l_{1-2} - l_{Turbine} = \{C_v(T_A - T_1) + p_A(v_2 - v_1)\} - C_v(T_2 - T_B).$$

The last term of the above equality shows that the work done within the air turbine increases the cycle efficiency by decreasing the total work. To realize this effect, the

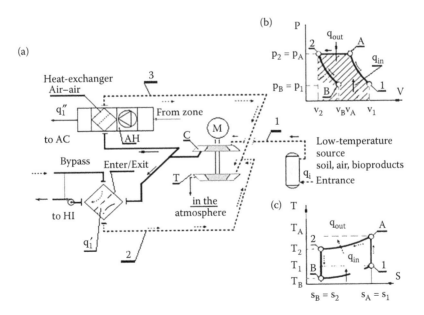

FIGURE 4.7
Air-compression heat pump, operating of the Brayton counter clockwise cycle: M—motor; C—air compressor; T—air turbine; HI—heating installation; AC—air conditioner; and AH—air handler of AC. (a) Equipment arrangement; (b) diagram of the mechanical work; and (c) diagram of the thermal work.

working fluid is cooled down but is still at high pressure (p_2). Then, it is introduced into the turbine after "release" of the regenerated energy $q_{Out} = q_1' + q_1''$ (into the heat exchangers of AC (q_1') and HI (q_1'') in this specific case). *The torque generated in the air turbine* T compensates a large part of the compressor resistance, although that the basic force needed to drive the turbo-compressor is generated by the engine (M). Thus, the external energy needed for compression during the initial process 1-A, is reduced.

Note that this is the process of practical *energy regeneration*.

The coefficient of performance of the air-compression heat pump takes the form

$$\eta_{T\Pi} = \frac{q_{Out}}{L_{BH}} = \frac{(q_1' + q_1'')}{L_{BH}}.$$

The regenerated heat q_{out} is utilized still at the turbo-compressor exit with temperature T_A. It is work-capable and can be used in various engineering projects of central heating as shown in Figure 4.7a.

The air heat pump aggregate has simple mechanics, and it is not "intricate" in operation. It is extremely efficient but its air ducts should be equipped with effective noise absorption material. The aggregate can be used in industry and ground transport.

4.2.3 Regeneration of Low Potential Thermal Energy Using a Vapor-Compression Heat Pump (Counterclockwise Rankine Cycle)

This technology is formally identical to the Brayton air-compression technology but the working fluid used here is evaporated by the low temperature energy source.[*] Figure 4.8a shows the scheme of connecting the devices into a vapor-compression heat pump, which operates in the counter clockwise Rankine cycle and is used for heating (its condensing heat exchangers are connected to a heating HI and air conditioning AC installations).

The working fluid (refrigerant) flowing through pipes, which are integrated to the components of the "cold" energy source, contains waste energy whose working capacity should be restored. Soil layers (at a depth of 3–5 m below the surface) play the role of a cold energy source shown in Figure 4.8a. It is found that their annual temperature is not affected by the change of the air temperature, and it stays positive (2–5°C). The integrated pipes, through which the working fluid is flowing, are manufactured from corrosion-resistant synthetic material and have high thermal conductivity (heat conducting plastics or polypropylene). Two options of pipe integration are shown in the scheme:

- Pipe integration within the building foundation slab (Figure 4.8a)
- Pipe integration within the structural components (building columns called "energy pylons" (Figure 4.8d)

The requirement is that the structural elements containing the pipes should be buried into the soil to a depth of at least 3–5 m below the surface, in order to provide steady

---

[*] The reason for using easily evaporating liquid instead of air is the same as that for employing a direct thermodynamic cycle. The type of the working fluid is chosen as depending on the expected temperature of phase transition which is different in different cases.

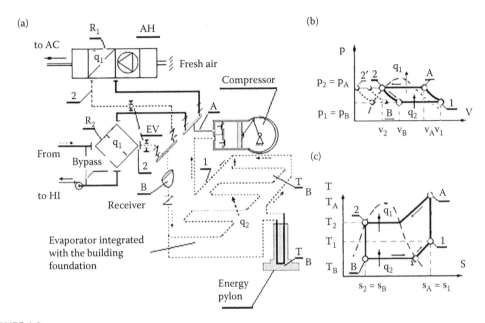

FIGURE 4.8

Vapor-compression heat pump, operating of the Renkine counter clockwise cycle: R_1 and R_2—heat exchangers; EV—expansion valve; HI—heating installation; AC—air conditioner; and AH—air handler of AC. (a) Axonometric scheme of the heat pump (HP) arrangement; (b) diagram of the mechanical work done within the Rankine cycle; and (c) heat diagram of the cycle.

conditions for the system and attain high energy efficiency. Control of the operation of such heat pumps starts at the receiver, where the liquid working fluid is collected.[*] Its parameters should correspond to those of point B in the state diagram (Figure 4.8b and c).

The liquid flows into the pipes of the evaporator under gravity and at sub-pressure (p_1) generated by the compressor. During flow, the liquid absorbs heat from the soil layers or from the steel fiber reinforced structure in which it is embedded. It evaporates along the isobar B-1 but its temperature remains constant until the required phase transition (at point 1′) is accomplished.

Then, temperature increases to point 1 in relation to the degree of steam superheating. The generated vapor has a temperature close to that of the cold energy source (i.e., <15°C).

Having the parameters of point 1, the cold vapor is drawn by the compressor, and pressure rises to p_A depending on the needs of the utilization systems. The hot working fluid leaves the compressor as "superheated vapor" with a temperature T_A. If waste but regenerated energy is used in HI or AC, temperature T_A does not usually exceed 40°C. In special cases, it can exceed 100°C.

If multistage compressors are used, the work done for energy regeneration is assessed by the expression

$$L_{Comp} = L_{Out} m^* z,$$

---

[*] Point B belongs to the state diagram and corresponds to a boiling liquid with wet steam fraction.

where

- m^*, kg/s—mass flux of the working fluid[*]
- z—number of compression stages

Hence, the power needed by the engine driving the compressor is found as

$$N_{eff} = k(L_{Comp}/\eta_M),$$

where

- η = 0.95–0.98—mechanical efficiency
- k = 1.15–1.20—coefficient of reserve power

Superheated vapor carrying the regenerated energy is supplied to the heat exchangers[†] of condensers R_1 and R_2—see Figure 4.8. There, energy q_1 is "extracted" from the working medium, and is used for technology needs or to heat the working fluid of AC and HI, as illustrated in Figure 4.8a. The superheated vapor itself is condensed at high pressure (p_2) along the isobar A-2, and it is transformed into boiling liquid at point 2, leaving the regenerative heat exchanger (point 2), as a high pressure liquid.

Assume that the task is to close the thermodynamic cycle and evaporate the working fluid under temperature T_B and at the low temperature evaporation surfaces embedded in the structure (the cold energy source). Then, the working fluid pressure should be reduced to its outflow value p_B at point B. That reduction can be performed in the expanding machine (its structure is similar to that of the piston compressor but it functions as a hydraulic engine[‡]). The typical vapor-compression cycle, however, uses an expansion valve (EV) (throttle-diaphragm EV) where pressure drops to the required value p_B by sudden expansion of the working fluid (throttling) along line 2-B—see Figure 4.8b and c.

Although that partial evaporation of the working fluid takes place during this process, thus decreasing the efficiency of the energy-absorption process developing at the evaporation surfaces, the throttle process is given preference due to its simplicity of application. The evaporation during throttling decreases if point 2 shifts to the left occupying position (2') via sub-cooling in the recuperative heat exchanger. After throttling, when the working fluid pressure reaches the prescribed value p_B, the liquid can be evaporated under temperature T_B and the cycle can be continuously carried out.

The example given is chosen regarding the operation of heating systems and AC of buildings which use soil and soil adjacent layers as CSs of the HP (the HS is the building itself). Those schemes are quite popular in countries with very cold climates. There, heat is "taken" from the soil and is transferred to the building heating system. Yet, in tropical

---

[*] Since power needed by AC and HI varies and depends on a number of random factors, the mass rate of the working fluid m* should be regulated in accordance with the current power sought. The regulation of m* takes place by choosing an appropriate compressor (scroll-compressors which regulate m* via rotations are suitable in this case).

[†] A wide variety of regenerating heat exchangers exists (for instant water/refrigerating agent, air/refrigerating aggregate, and soil/refrigerating aggregate). This broadens the abilities of design of stem compressor and air-compressor pumps of various applications. See more in Radermacher R. and Y. Hwang (2005).

[‡] The expanding machine included in the Rankine cycle functions as the air turbine participating in the Brayton cycle.

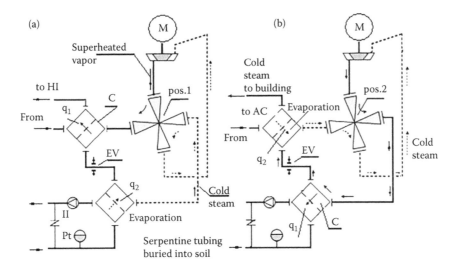

FIGURE 4.9
Reversive heat pump technology with four-way valve: (a) position 1—building heating and (b) position 2—building cooling.

and equatorial areas, buildings are used as cold HP sources while the soil layers beneath are used as heat sinks if no other options (as running or lake water, for instance) are available.

Countries with four-season continental climate buildings need heating as well as cooling. Hence, a problem of universal use of regenerative heat exchangers emerges, so that one could realize a combined regime of building operation using the very same equipment. This can be done by using the so-called "reversible heat pumps" (Figure 4.9).

All components of the heat pump found in the designs just described a remain here. Only a four-way valve with two positions control is added to the scheme as a new element. Its incorporation allows for the change of the function of the two regeneration heat exchangers R_1 and R_2, and the contacts of the heat pump with the cold and hot energy sources are reversed.

This specificity is illustrated by the two schemes shown in Figure 4.9a and b which are totally identical but with different steam distribution due to the two different positions of the four-way valve.

When the four-way valve occupies position 1, the working fluid leaves the compressor as superheated vapor and is directed to the HP condenser. Yet, it follows the allowed direction and enters the heat exchanger R_1 connected to the heating loop of the building. It releases heat q_1 and is liquefied. The condenser high pressure (p_2) decreases owing to action of the EV. Then, the working fluid is injected into the heat exchanger R_2 connected to the CS (the soil around the building). There, the working fluid evaporates taking heat q_2 out of the soil. These cold vapors leave R_2 and are drawn by the compressor through the "allowed" channel to repeat the inverse heat pump cycle.

Thus, the heat pump takes the low-potential heat out of the CS, transforms energy state parameters (temperature and pressure), regenerates energy via external mechanical work done by the motor (M), and finally directs it to the heating loop of the building.

When the four-position valve is in position 2, the superheated vapor flowing in the new *"allowed channel"* is directed to the regeneration heat exchanger R_2 instead of R_1. Since R_2 is linked to a low temperature medium (at T_1), the superheated vapor easily releases the latent heat q_1 of the phase transformation. It is then liquefied after throttling and reduction of pressure to p_B, while the working fluid flows through the regeneration heat exchanger R_1 which now serves as an evaporator. The working fluid evaporates taking heat q_2 from the "area of comfort" and cooling it. The heat pump machine operating on this cycle uses the "area of comfort" as a CS, while the low-potential medium—is the HS. Thus, it can be used as a component of building AC.

The capacity of steam- and air-compressor heat pumps is predicted on the basis of the expected power, needed by the systems that draw energy from their condensers (the recuperative heat exchangers) at extreme heat loads.[*] The appropriate choice of a working fluid (refrigerant) is critical for the efficient operation of vapor-compression heat pumps, since the coefficient the performance depends on the type of working fluid used.

We may conclude that together with vapor- and air-compression heat pumps, other types of heat pumps also exist. These are the absorption pumps where binary chemical mixtures are used to absorb heat from the CS and subsequently release it to the HS. These heat pumps are subject of specialized teaching courses on refrigeration equipment.

4.3 Recuperation Technologies for Thermal Waste and Environmental Energy

The second large group of technologies for energy conversion is that of technologies for "energy recuperation."[†] A detailed list of these technologies is schematically shown in Figure 4.1. Disregarding the origin of the original primary energy sources (industrial, daily, transport, technological energy waste, or environmental energy), one can classify the recuperation technologies as

- Direct ones (similar to those in Section 3.1)
- Cogeneration technologies
- Indirect ones (as those in Section 3.2)

The second and the third technologies can bring about at least two energy conversions.

Initially, energy waste (with low temperature and pressure) excites mechanical motion[‡] of macro-bodies equipped with electromagnetic solenoids (coils). Then, the mechanical motion undergoes a second energy conversion, and it is transformed into electric current (current of electric charges—electrons and ions) by means of "electromagnetic induction." The devices converting the mechanical energy into electrical energy are called "electrical generators." They have undergone significant evolution as shown in Figure 4.10. It is assumed that the first electric generator (Figure 4.10a) was designed

---

[*] The cooling load of AC costs of heat released by people, lighting, and technological equipment, as well as heat due to the environment. The heating load is the difference between supplied and lost heat within an occupied area under extreme environmental conditions.

[†] By definition, the recuperation is based on the change of the energy form. We discuss in the present chapter recuperation which is used to generate electricity, only.

[‡] The energy carrier, that is, flowing fluid (steam, combustion products) transfers its kinetic energy to the rotating or translating macro components of operational rotors or pistons of external heat engines (engines of Stirling, John Barber, and Larkins), stimulating their motion in a magnetic field.

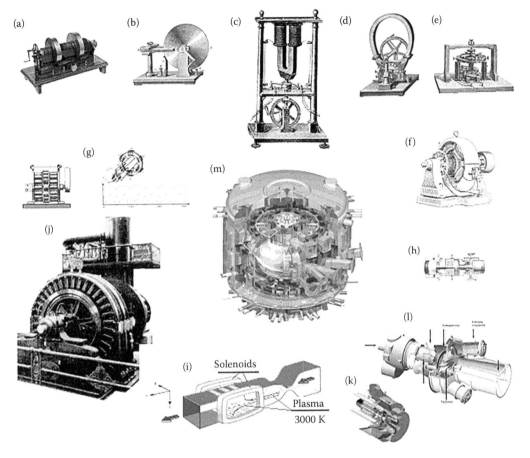

FIGURE 4.10
Evolution of the electric generators: (a) Jedlik's generator; (b) Faraday's one-pole generator; (c) generator of Hippolyte Pixii; (d) Gramme ring; (e) Pacinotti; (f) multipole generator; (h) linear generator; (g) and (j) Tesla's generators; (i) cold plasma MHG; (k) cogenerator; (l) cogenerator with gas micro-turbine; and (m) hot plasma MHG.

in 1827 by the Hungarian inventor A. Jedlik. This occurred 4 years prior to Michael Faraday's discovery of electromagnetic induction in 1831 and the demonstration of Faraday's single-pole disk (Figure 4.10b). Jedlik's prototype generated electric power of 3 kW (9 V/310 A). Yet, it was not realized on the market, since it was not patented till 1850 (see Figure 4.10a).

Faraday's generator (Figure 4.10b) was modified and improved by Hippolyte Pixii in 1832 (Figure 4.10c) and E. Siemens who designed the first generator of direct current applicable in industry in 1867. The Siemens "dynamo," being universally recognized and realized on the market, was used with tremendous success till 1891 when the first generator for alternating current was designed (Figure 4.10j) by the great Nikola Tesla (1856–1943).*

---

* Nikola Tesla's father, Milutin Tesla, was an orthodox priest in the Latin eparchy of the Constantinople Patriarchate. According to some sources, he came from Berkovitsa, a town in Bulgaria. His parish was founded for the needs of Polish orthodox officers in the service of the Austrian Empire in the village of Smiljan, 7 km northwest of Gospich—a town in today's Croatia.

This was considered to be the crucial moment of the industrial revolution in Europe and North America, since it laid the foundations of total electrification.

Developing a generator for alternating current and electric transformer prototypes, N. Tesla laid the foundations of long distance energy transfer and "medium" and "high" voltage grids, building the energy infrastructure of a certain territory (national, continental, or urban). Parallel to the rotating generators, two more "induction" concepts were developed: those of linear (Figure 4.10h) and magnetic-hydrodynamic generators (MHGs) (Figure 4.10i and m). Linear generators convert the reciprocating mechanical motion of macro-solenoids into electric current. They have not gained popularity but they can be successfully used in *cogeneration modules* or as *recuperators* of waste energy in industrial technologies, together with Stirling engines.

The MHGs were initially built in the late 1950s and had small power approximately 12 kW. Later, they were improved to operate using low temperature plasma produced by natural gas combustion. Their operational temperature ranged up to 3000 K, and the highest power was 25 MW (YO2-1965) and 32 MW (Mark V—the late 1960s). They were not realized on the market, since their efficiency was lower that of the Brayton gas-turbine generators, considering use of the same fuel and operational temperature.

The main recuperation technologies are based on the already discussed thermal-mechanical and thermal-electrical technologies (Sections 4.2 and 4.3). To avoid repetition, we shall not analyze them additionally in detail. In what follows we shall discuss those technologies that have not been mentioned above. These are namely the osmotic, hydro-aerodynamic, and optic-concentrator-regeneration technologies.

4.3.1 Osmotic-Absorption Technology

Although osmotic technology did not become popular until 2009 when the first electro-productive aggregate was activated in Norway, its principle of operation was known for three centuries. The so-called "osmotic engine" was invented by the great hydromechanic Daniel Bernoulli in 1700. Its structure and operational principles are shown in Figure 4.11a. It operates without the introduction of primary energy carriers, such as biomass, coal, dressed uranium, or natural gas. Purely chemical reactions do not take place in the engine and hence, no hydrocarbon or nitrogen oxides are released. It is often taken to be a prototype of a "perpetual motion machine." The engine structure is quite simple, consisting of

- Two vertical cylinders 1 and 3 filled with liquid
- Two horizontal bypass pipes A and B
- Water turbine 4 (Figure 4.11a)

A semipermeable membrane 2^* is mounted in bypass A—see Figure 4.11a, which is the only "hi-tech" element in the Bernoulli engine. To operate, the osmotic engine needs only a water solution of sodium chloride or other salt, and it is poured into cylinder 1. Membrane 2 operates as a filter of the salt water keeping the NaCl ions in cylinder 1. Thus, it creates nonequilibrium in the system— only clear water passes through it.

---

* Two types of membranes are used

  - CAB, manufactured from cellulose acetate (by thin film casting)
  - CPA, manufactured from polyester cloth coated with polyamide (polysulfon lader)

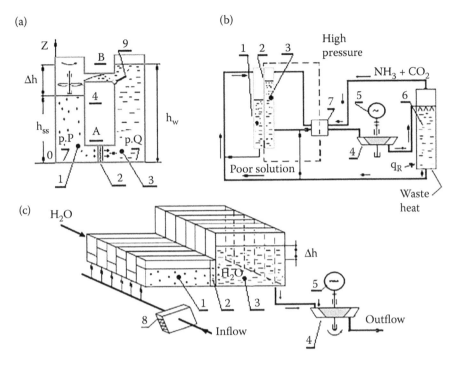

FIGURE 4.11
Principal scheme of an osmotic engine: (a) Bernoulli engine; (b) osmotic recuperation installation with AOM; and (c) scheme of an electric power station: (1) salt solution; (2) membrane; (3) water; (4) water turbine; (5) AOM generator; (6) AOM reactor; (7) mixer; (8) filter; and (9) gate.

Both cylinders accumulate liquids with different density: cylinder 1 contains a salt solution while cylinder 2—clear water. This is *the first specific feature* of the Bernoulli engine. (Water density under normal conditions is $\rho_{H_2O} = 1000$ kg/m³.) Density of the salt solution depends on the concentration of the sodium chloride. Salt water density in the Red Sea for instance is about 1090 kg/m³. *The second specific feature* in this case is that *the clear water column* is higher than the other column. The next paragraph gives the reasons for and the consequences of this phenomenon.

It is known that under gravity, surfaces of constant pressure (isobaric surfaces) are horizontal planes perpendicular to the gravity acceleration (see Figure 4.11a where they are perpendicular to the axis Oz). The following equation for pressure holds for all isobaric planes, including that coinciding with the bottom of bypass 1 (plane I-I):

$$p_P = p_Q = p_{I-I} = const.$$

Here points P and Q lie in the isobaric plane but still experience the weight of two columns of fluids with different density.

As is known from mechanics, the following link between pressure at those points, fluid density ρ and column height h exists (g is gravity acceleration):

- Pressure at point P: $p_P = g \cdot \rho_{P-p} \cdot h_{P-p}$
- Pressure at point Q: $p_Q = g \cdot \rho_{H_2O} \cdot h_{H_2O}$

On the other hand, points P and Q lie in the isobaric plane, that is,

$$\rho_{P-p} \cdot h_{P-p} = \rho_{H_2O} \cdot h_{H_2O}.$$

Thus, the height of the water column reads

$$h_{H_2O} = h_{P-p} \frac{\rho_{P-p}}{\rho_{H_2O}}, \tag{4.7}$$

but $(\rho_{P-p} > \rho_{H_2O})$, whereas $h_{H_2O} > h_{P-p}$. Hence, the height of the water column h_{H_2O} is larger than that of the column of the salt solution h_{P-p} due to the higher density of the latter. The positive difference Δh between the two columns is called "piezometric height" or "displacement." It generates the hydrodynamic head p_H on the normally closed gate 9—see Figure 4.11a, and it is set forth by

$$p_H = \rho_{H_2O} \times \Delta h = \rho_{H_2O} \times h_{P-p} \left(\frac{\rho_{P-p}}{\rho_{H_2O}} - 1 \right). \tag{4.8}$$

Putting the expression (4.8) for h_{H_2O} in the difference $\Delta h = h_{H_2O} - h_{P-p}$ we find

$$\Delta h = h_{H_2O} - h_{P-p} = h_{P-p} \left(\frac{\rho_{P-p}}{\rho_{H_2O}} - 1 \right).$$

Now, if gate 9 opens, water flows out of cylinder 3 through bypass B and into cylinder 1 at hydrostatic pressure p_H. The mounted water turbine converts the kinetic energy of the clear water flowing from cylinder 3 to cylinder 1 into useful mechanical work of the shaft. For instance, if the turbine (see Figure 4.12) is combined with an electric generator,

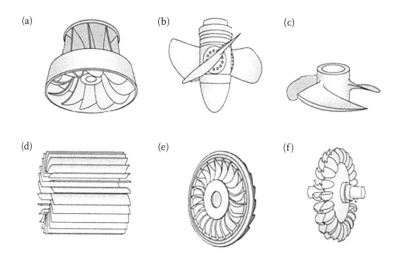

FIGURE 4.12
Water turbines used in osmotic technologies: (a) Francisturbine; (b) Kaplan turbine; (c) turbine with a propeller with fixed blades; (d) turbine with a diametric propeller; (e) turbo-propeller; and (f) Pelton turbine.

the turbine kinetic energy will be converted into electric power. The regulation of the operation of the Bernoulli engine is performed by gate 9. It is such that the flux to cylinder 1 would be equal to that passing through the membrane, keeping the height of the water column h_{H_2O} constant.

Remaining in the "continuity" condition (see Equation 1.16), the Bernoulli osmotic engine can operate continuously and "for free." Even though its principle of operation was described long ago, it has not been industrially applied on large scale. The reason is the low power per membrane unit area (the classical scheme allows the "squeezing" of 1 W/m² out of an osmotic cell). The power of the osmotic engine* is increased:

- By the increase of the mass rate of the "clear" fluid (using for example a series of osmotic cells, simultaneously operating in parallel—Figure 4.11c or
- By the increase of the hydrostatic head magnitude (special solutions are used in this case admitting manipulation of the operational pressure, as is in the absorption osmotic machine—see Figure 4.11b.

The use of thick liquids such as saccharose, starch, proteins, etc. may result in the generation of specific power of 3–5 W/m².

A significantly more effective approach to increase the power of the hydrostatic pressure and the engine specific power, respectively, is the use of the absorption osmotic engine (Figure 4.11b). Those machines justify our consideration of the osmotic technology, since they operate using waste heat. The absorption osmotic systems convert the waste heat of a certain technological process into electrical energy, and they are quite suitable for exploitation as industrial recuperators. The working fluid is sea water with ammonium (NH_3) and carbon dioxide (CO_2) as thermolytic admixtures prepared in mixer 7 in Figure 4.11b. The dressed mixture expands in turbine 4 converting the hydrostatic energy accumulated in the osmotic cell into kinetic energy. Then, it is collected in the reactor 6 of the absorption machine. The dressed solution is heated over 20°C owing to the waste heat, and the refrigerating agent (the thermolytic admixture) evaporates, leaving the salt solution (the absorbent) through the reactor upper section. Since the salt solution is heavier than the high-concentration mixture, it flows into the osmotic cell through the vertical cylinder 1. The semipermeable membrane 2 filtrates the low-concentration solution and the clear water released mixes with the thermolytic component and is accumulated in the second cylinder 3 as well—see Figure 4.11c.

Then, the hydrostatic pressure p_h increases due to the head of the gas component. At high hydrostatic pressure, the high-concentration solution flows in turbine 4 which is coupled to the electric generator 5. The solution thus utilized leaves the turbine and enters reactor 6, where it is again heated to $T > 20°C$ to repeat the cycle.

The energy efficiency of the osmotic absorption technology is assessed via the conversion coefficient, and it is within ranges 50%–80%, while the specific power of recuperation amounts to 200 W/m². Thus, it can be really employed in modern design.

---

* The turbine hydraulic power N_{Typ} is the product between the hydrostatic pressure p_H (the head of a water column with height Δh) and the flux of the working fluid entering the turbine.

4.3.2 Hydrodynamic Recuperation Technologies[*]

Hydrodynamic technologies are the oldest known power technologies. They have the longest history[†] being applied in daily life, transport, and industry. The authentic natural energy source for them is solar energy which is converted into air circulation[‡] and water natural circulation.

The basic component of the energy equipment of a hydrodynamic technology is the propeller (a rotating macro-body or a rotor). Its task is to convert the hydrodynamic power of the flowing fluid into kinetic energy. Regardless of the type of the energy carrier and variety of design, propellers are

- With vertical axes (Figure 4.13a)
- With horizontal axes (Figure 4.13b)

A common feature of those devices is the physical mechanisms of interaction between the working fluid and the propeller blades (the stream-lined bodies).[§] The aerodynamic force $\vec{R}_A$, which is a measure of this interaction, emerges and acts on the lateral surface of

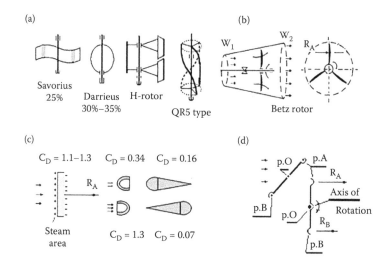

FIGURE 4.13
Types of wind propellers: (a) vertical axes (Savonius, Darrieus, H-rotor, Quietrevolution QR5 Gorlov type); (b) horizontal axes (Albert Betz); (c) propeller wings (stem area; cup- and wedge-shaped blades); and (d) forces, accelerations, and moments.

---

[*] The difference between hydrodynamic recuperation technology (HDRT) and Magneto hydrodynamic technology (MHDT) lies in the state of the working fluid. Regarding the first type, molecules are not ionized and there are no electrodynamic interactions. As for the second type, the fluid is ionized.

[†] The oldest design of a windmill dates from the second century BC in Persia. The use of windmills in Europe did not start until the 14th century in Holland. The first wind electric power station was built in 1887 by the Scottish inventor J. Bluth.

[‡] The Earth's atmosphere is a component of a natural engine with negligible efficiency (<2%). The power of Earth air fluxes amounts to about 3,5 TW.

[§] The emergence of aerodynamic elevation forces is studied in courses on Fluid Mechanics. In brief, if a fluid contacts a solid wall, an area of increased external pressure emerges (i.e., a stem area—Figure 2.43c) and forces acting on the solid body surface increase. The resultant force R_a elevates the body and generates acceleration a_a along its directrix.

the solid body (propeller blade) and depends on the fluid velocity (w_B), fluid density (ρ), and body shape (stream-line form—C_D) and "active" cross section area (A_0).

The following formula is due to Bernoulli, and it is used to calculate R_A:

$$R_A = p_n \cdot A_0 = 0.5 \rho w_B^2 A_0 \cdot C_D,$$

where C_D is the coefficient of body aerodynamic resistance (see Figure 4.13g).

Consider the operation of the most popular rotor—the cup-shaped propeller (Figure 4.13c). To start rotating, it should be "deprived" of two degrees of freedom (two plane translations). Then, the aerodynamic force will generate torque about a point fixed as a center of propeller rotation (point 0, for instance).

Regarding the direction of fluid flow, the propeller will rotate clockwise since the two forces ($\vec{R}_A$ and $\vec{R}_b$) generate opposite torques with different magnitude. This is due to the different values of C_D (the coefficient of aerodynamic resistance) $\to M_{R_A} > M_{R_B}$ (see Figure 4.13c and d).

The efficiency of the cup-shaped propeller is very low and within ranges the 7.5%–16%. The operation of the ideal propeller of Betz (Figure 4.13b) is significantly more efficient, and its efficiency reaches 59.3%. This is the limit when aerodynamic interaction between the working fluid and the propeller takes place. Hence, a Betz propeller is used as an energy standard for real propellers, whose efficiency is within ranges 40%–50%. The available aerodynamic power is a function of the current velocity of the fluid (W_B), and it is predicted by the expression[*]

$$N = \dot{E} = \frac{d}{dt}(0.5\, mw_B^2) = 0.5\, \dot{m}\, V^2 = 0.5\, \rho A_0 w_B^3,$$

while propeller power N_T is assessed by

$$N_T = M_0 \omega_0 = 0.5 \cdot c_N \cdot \rho \cdot A_0 w_B^3,$$

where W_B is wind velocity; c_N is power coefficient depending on the velocity ratio $\lambda = u/W_B$ specified in Figure 4.14.

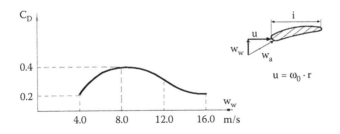

FIGURE 4.14
Power coefficient of wind turbines.

---

[*] At velocity 30 m/s and air density $\rho = 1.245$ kg/m³, wind power reaches 16.6 kW/m². The same power can be attained if density is 1000 kg/m³ and velocity is 3 m/s, but using another working fluid.

(a)   (b)

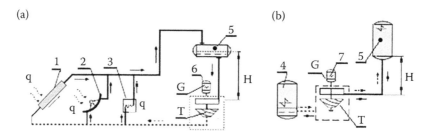

FIGURE 4.15

Hydrodynamic recuperations: (a) thermo-siphon recuperation and (b) pumping-heat storage recuperation: (1) flat solar collector; (2) solar concentrator; (3) regenerative heat exchanger; (4) lower fluid storage; (5) top heat storage; (6) turbine/pump aggregate; and (7) reversible aggregate.

Hydrodynamic technologies are applied in schemes using renewable sources—wind power and the power of rivers and small creeks. Figure 4.15 shows examples of their usage for the recuperation of waste heat or for the generation of cheap night energy. The first of them (Figure 4.15a) shows industrial waste heat q_R utilized by the working fluid via heat exchanger 3. The fluid is heated along a regeneration cycle (see Figure 4.3a), its temperature T increases and its density decreases. Thus, ascending convective motion directed from the heat exchanger 3 to the accumulator 5 is initiated. The accumulator is mounted at height H, m over the turbine-generator aggregate[*] that generates electricity.

Two types of solar collectors (plane, with, and without vacuum pipes or collectors with solar concentrators) are alternative devices for heating the working fluid within the regeneration heat exchanger. The mounting height H of accumulator 5 depends on the building architecture and on the eventual possibility to fix the accumulator at the building highest point. The second scheme (Figure 4.15b) is a replica of a pump-accumulation electric power station (PAEPS). Such a scheme can be used in industrial buildings supplied with a one and two-tariff system of charging for the consumed electrical energy.

The difference between the proposed pumping-accumulation scheme and the scheme of thermo-siphon regeneration is in that besides the removal of the regenerative heat exchanger, the turbine-generation aggregate should be of reversible type and be able to operate as a pumping device.

During periods of high cost electricity (from 10 a.m. to 2 h p.m.), the aggregate should operate as a generator producing electrical energy for its own needs or returning electricity to the network (using the potential energy stored into the top heat storage 5). During the time span of low cost electricity (0–6 h a.m., for instance), the aggregate 7 should operate reversely as a pump drawing the working fluid from the lower fluid storage 4, pumping and elevating it to height H and supplying it to the top heat storage 5. This system could operate on 24-hour-a-day cycles, and it saves money when conventional energy costs are high. The payback time of the initial investments depends on the ratio between low and high energy costs and on the system capacity.

---

[*] Small water turbines are used, type "Caplan," "Francis," "Pelton," "turbo-centrifugal," "Crossflow," or with a "fixed propeller."

4.3.3 Radiative-Optical Technologies Concentrating Solar Energy

The Sun's vital importance* was known since ancient times. Ancient people worshipped the Sun (it was god Amun Ra for the Egyptians, god Tangra for the proto-Bulgarians and other Palmarians, god Apollo for the Greeks, etc.). Those beliefs were owing to the hidden physics of solar energy, whose mystery grasped the human imagination until the mid-19th century, despite the treatise of Isaac Newton and other physicists on this matter.

Michael Faraday was the first to formulate a hypothesis that solar energy was a set of electromagnetic waves. It was confirmed by a number of physical experiments proving the energy wave character. However, other experiments proved its mass properties thus contradicting the previous experimental evidence.[†]

The riddle of the nature of solar energy was one of the toughest problems that physics faced. Thanks to the great German physicist Max Plank and to his quantum theory, modern science has gained profound knowledge on the nature of the solar energy and on its wave-mass properties. Hence, owing to the nuclear reactions occurring in the Sun (see Figure 4.16), fluxes of photons and other energy particles outflow and spread in the surrounding space. The density of the energy flux at the Sun surface is 63.11 kW/m² (the Sun emits about 400 EJ per year). Only a small energy fraction reaches the Earth. Photons are the basic elementary particles of the solar energy that reach the Earth. The density of a photon flux meeting a surface which is perpendicular to the solar rays is assumed to be constant, equal to 1390 W/m².[‡] That density has been measured outside the atmosphere (in the exosphere).[§]

According to modern physics, the flux of solar energy is composed of solar photons, which are quasi-particles of the class of bosons, since they possess a whole-number spin. A basic property of the boson is that it spreads as a "standing wave" with angular frequency $\omega_0 = 2\pi v$ (V-frequency) which is a constant integer and does not vary during photon motion. Hence, the photon energy according to Max Planck is also constant

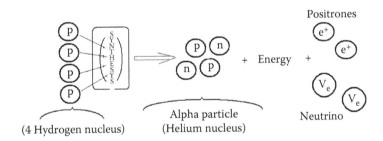

FIGURE 4.16
Result of the nuclear synthesis is hydrogen conversion into helium, releasing positrons e+, neutrons v_e, and photons (energy).

---

* The Sun is a star whose age is about 5×10^9 years, and it is in the middle of its "life" cycle. The sun surface temperature is 5777 K (15×10^6 K at the Sun center). The sun is composed of 80% hydrogen and 20% helium (as well as 0.1% other elements), and its mass is 1.989×10^{30} kg.

† It has been even proved that solar energy affects surfaces that are normal to its directrix.

‡ Due to the variation of the distance between the Sun and the Earth, the solar "constant" varies between 1325 and 1420 W/m².

§ The energy of direct radiation damps when penetrating the atmosphere, and it varies within comparatively wide limits (17%–60%). That phenomenon is explained by mass effects and by the effects of Raleigh and May.

$$\langle e_{pk} \rangle = \hbar x \omega_0 = 1.05 \times 10^{-34} \, \omega_0, \, eV,$$

where $\hbar = h / 2\pi = 1.05 \times 10^{-34}$ Js is the Planck constant. Various numbers $\bar{n}(\omega_0)$ of photons (bosons) can be found within a standing wavelength with angular frequency ω_0 (frequency mode). Here, $\bar{n}(\omega_0)$ is called the photon distribution function, which is identical with that valid for all bosons. It is specified by Bose's function

$$\bar{n}(\omega_0) = \frac{1}{\exp(\beta \hbar \omega_0) - 1},$$

also known as the Einstein–Bose function.

The exponent power β (Lagrange multiplier) takes values inversely proportional to the absolute temperature T. The link is specified by the equation

$$\beta = \left(k_B T \right)^{-1},$$

where k_B is the Boltzmann constant already discussed. Thus, the number of photons in a certain energy orbital (the frequency mode) will depend on the temperature of the "photon gas"—T

$$\bar{n}(\omega_0) = \frac{1}{\exp(\hbar \omega_0 / k_B T) - 1}.$$

It will increase with the increase in the temperature T of the photon "gas." It is found that with temperature decreasing photons leave the high-frequency orbitals and pass to low-frequency ones while the distribution function takes another form.

As said, the Sun core temperature is about 15×10^6 K, while that of its surface is 5777 K.[*] This is decisive for the type of the photon distribution function $\bar{n}(\omega_0)$. It turns out that under the Sun's surface temperature, photons reaching the Earth surface in the form of direct radiation are divided into three groups:

- A group of high energy photons with wave length $\lambda < 350$ nm and energy of $e_{Ph} > 3.1$ eV (they carry about 7% of the total energy)
- A group of visible photons with wave length 350 nm $> \lambda < 700$ nm and energy $3.1 > e_{Ph} > 1.77$ eV (with energy share of about 47%)
- A group of infrared photons with $\lambda > 700$ nm and energy $e_{Ph} \leq 1.75$ eV (with energy share of 46%)

The rest of the photons scatter owing to the air molecules including the water vapor molecules and the aerosols.

Besides the photons of solar radiation (direct and scattered), photons of reflected solar flux also participate in the energy balance of the absorbing solar surfaces (panels). Yet we shall consider only photons of the first and third group. This is due to the circumstance

---

[*] It is assumed that the temperature of the photon "gas" emitted by the Sun is about 6000 K.

that there are only two types of physical devices that can change the qualitative characteristics of the photon flux (gas), namely:

- Concentrators[*]
- Reflectors

Hence, it seems useful to systematize them in two groups of technologies improving the photon flux quality:

- Technologies of the reflecting geometry
- Technologies of the geometrical optics

They will be considered in what follows.

4.3.3.1 Technology of the Reflecting Geometry

Here, solar photons that enter the area of the energy convertor are concentrated on a small absorption surface by means of solar mirror surfaces. Thus, a multiple increase of their flux density over the natural value of 1395 W/m² is obtained.

Two reflection schemes "collecting" photons are known:

- A scheme using mirror trackers—Figure 4.17
- A scheme using tracing rotating bands—Figure 4.20

The presented "reflection" technology uses discrete solar-reflective surfaces 1, and the area of each of them is 4–6 m². They are mounted on manipulators with an open kinematic scheme called trackers. The mirror surfaces are positioned under control of a "sun-tracing"

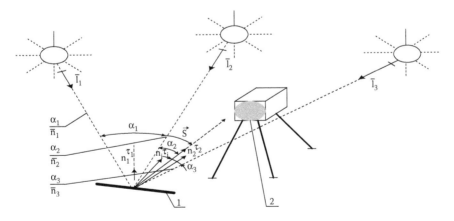

FIGURE 4.17
Mirror trackers with a tower receiver: (1) tracing mirror-reflecting surface and (2) fixed "central" receiver (central heliostatic tower).

---

[*] Although that the two devices are known since ancient times, modern technologies based on the achievements of electrochemistry, vacuum technology, and tracing system automation justify their consideration in the present expose.

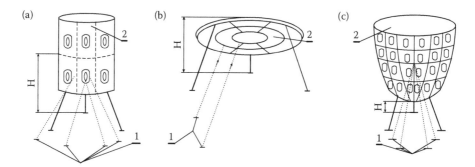

FIGURE 4.18
Fixed central absorber—a heliostatic tower: (a) cylindrical receiver covered with Fresnel lenses; (b) oval central receiver; and (c) parabolic central receiver: (1) reflecting surfaces and (2) heliostatic tower.

system, and the photons reflected by each discrete surface fall on the central absorbing surface 2 in the "heliostatic" tower (Figure 4.17). Figure 4.18 shows the used fixed absorbers schematically.

Large Genei interlocking solar concentrators (with diameter of up to 10 m) are used in small solar systems with power not larger than 40 kW. Note that 10–12 reflection planes are mounted on them so that the reflected solar photons would concentrate within a small control volume (common focal center—F), where the absorbing surface is positioned—see Figure 4.19. A large concentration of solar energy, 10 times exceeding the normal value, is attained in the area around the focal center F. This effect yields increase of the temperature of the photon gas up to several hundred thousand degrees (150,000–180,000°C), while gas energy is absorbed by a fluid flowing in the pipe absorber 2.

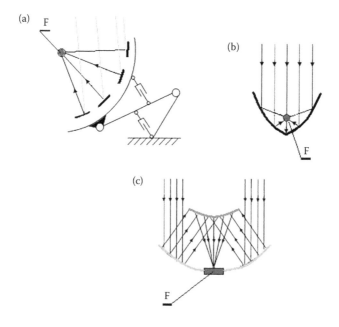

FIGURE 4.19
System of parabolic mirrors (dish system): (a) Genei interlocking solar concentrator; (b) point focus parabolic concave mirror; and (c) Cassegrain reflector.

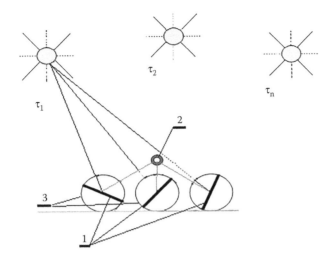

FIGURE 4.20
Linear Fresnel solar consentator: (1) linear Fresnel reflectors; (2) fixed pipe absorber; and (3) tracing disks.

The location of the mirror planes follows the contour of different second power surfaces (paraboloid, hyperboloid, ellipsoid, sphere, etc.). Each concentration of the photon flux can also be attained by a combination of concave and convex mirror surfaces as is in the Cassegrain reflector[*] (Figure 4.19c). A concave hyperboloid cylinder combined with a spherical convex cylinder, or a hyperboloid convex cylinder combined with an ellipsoidal convex reflector cylinder, is used for this purpose.

Those systems for "catching" the Sun's radiation are known as "systems of parabolic mirrors (dish systems)," and they are used combined with the Stirling electrogenerators, which are mounted in the neighborhood of the concentrator focal center. Other reflection systems, concentrating the solar energy flux, are manufactured using linear Fresel reflectors, which in their turn are plane (Figure 4.20) and parabolic (Figure 4.21).

When the system consists of narrow linear Fresel reflectors (item 1 in Figure 4.20), they are fixed along the diameter of disks rotating about their axes—position 3 (rotors with diameter equal to the band width). The disks of each band rotate such that the reflected Sun's rays would fall on the fixed pipe absorber 2. Thus, it is irradiated by fluxes reflected by several bands (three in our case), with which it operates jointly. In cases of linear parabolic reflectors 2—Figure 4.21, the tube absorber 1 is fixed to the parabolic mirror (with a shape of a trough), and it is positioned along the mirror focal axis.

4.3.3.2 Technology of the Geometrical Optics[†]

In recent decades, geometrical optics found a new application in solar thermal and photovoltaic technologies. This new trend called "panel optical light conductor" is in fact a technique for the concentration of the flux of solar photons falling on a small illuminated

---

[*] Laurent Cassegrain (1672) proposed a reflector consisting of a concave parabolic reflector and a convex hyperbolic reflector.
[†] Optical lenses were known since ancient times (in Persia). The theory of the geometrical optics is set forth in the tractates (11th century AD) and Ibn al Haytham (12th century AD).

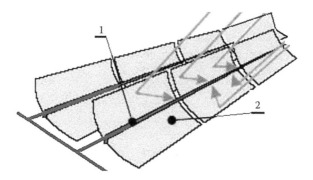

FIGURE 4.21
Parabolic trough reflectors: (1) fixed absorber tube and (2) linear parabolic reflector.

surface,* using optical lenses. In contrast to the classical applications of geometrical optics, solar technologies need larger diaphragms and shorter focal distance. The conventional lenses (Figure 4.22a) are bulky, thus increasing the equipment dimensions and weight.

The high requirements of the solar schemes can be satisfied by the Fresnel lenses[†]—Figure 4.22b, which are narrow, light, and cheap. Polycarbonate gel or active composite materials, instead of high-quality glass, are used in solar technologies. A decrease of lens volume is attained using concentric, ring-shaped sections along the periphery called Fresnel areas, which are a set of round prisms.

In thin lenses, the number of Fresnel areas grows and they can be used with PV cells (see Figure 4.23). Lenses with magnification 500:1 are used in solar systems while the temperature of the concentrated photon flux reaches 350,000–500,000 K.

Note in conclusion that the development of regeneration and recuperation technologies and their implementation in industry and structural engineering follow the general tendencies of "big" energetics, that is, the need for technological equipment with improved efficiency but with reasonable time of investment payback. The successful management of investments when employing these technologies requires the design

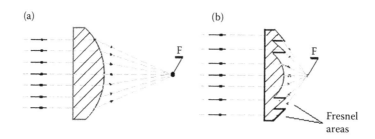

FIGURE 4.22
Concentrating lenses: (a) lens with a long focal distance and (b) the Fresnel lens has a large diaphragm and a short focal distance.

---

* Increase of the efficiency of the photovoltaic cells to 40%.
† George Leclerc proposed and Augustin Jean Fresnel designed in 1823 a thin lens to be applied in lighthouses.

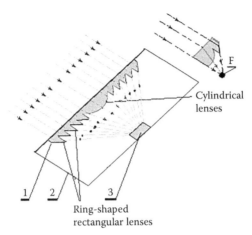

FIGURE 4.23
Photovoltaic cell with a Fresnel lens: (1) Fresnel lens; (2) carrying structure; and (3) miniature photovoltaic cell.

of appropriate instruments, including official rating tables with indices and classes of energy efficiency.

4.4 Control Questions

1. What is the difference between the two types of energy conversion: regeneration and recuperation?
2. What is the role of entropy in the assessment of energy quality?
3. How does a TDS de-energization proceeds on micro- and macro-level?
4. How can one calculate entropy change?
5. Is entropy increase always symptomatic for system de-energization?
6. Give examples of thermodynamic processes where, despite entropy increases, a TDS displays increased energy potential.
7. What are the advantages of the methods of the statistical thermodynamics used to analyze processes of energy conversion?
8. What is the difference between the collective macrostate and the minoritarian microstates?
9. How does the change of the number of macrostates affect the recorded values of the system parameters?
10. How does a TDS de-energization proceed on micro-level?
11. How do energy-exchange processes proceed in a noninsulated TDS?
12. How should one define the index of energy efficiency of TDS de-energization?
13. What is the physical interpretation of the term "exergy"?
14. How does the index of energy efficiency change during TDS de-energization?

15. What is the direction of index change during TDS energization?
16. Specify the components (processes) of a regeneration cycle.
17. Give the basic types of regeneration devices.
18. Which are the physical quantities that the index of regenerator energy efficiency deepens on?
19. How does a heat pump operate?
20. Using a heat pump, is it possible to generate energy whose amount is larger than the amount of energy "extracted" from the CS?
21. Which are the two types of the most popular heat pump technologies?
22. Which heat pump employs the Rankine cycle? Which heat pump employs the Brayton cycle?
23. Specify at least three differences between the equipment of air-compressor heat pumps and that of steam-compressor heat pumps.
24. Which is the basic difference between the two counterclockwise thermodynamic cycles (those of Rankine and Brayton) regarding the behavior of the working fluid?
25. What component should one add to the scheme to adapt a heat pump to a reversive operational regime?
26. Specify the basic groups of recuperation technologies where change of the energy form takes place.
27. What is the difference between direct and indirect recuperation technologies?
28. Describe the thermoelectric recuperation technology used in a plant for refuse combustion (using a gas-turbine electrogenerator).
29. Describe the thermoelectric recuperation technology applied in a steam-turbine technology used in a dung-hill depot.
30. What is the equipment of magnetohydrodynamic installations used in refuse-combustion plants?
31. Describe the principle of operation of the osmotic engine.
32. How can one increase the power of the osmotic engine?
33. What is the value of the specific power of an osmotic engine?
34. Design a scheme for the use of waste heat from a hotel sewer system equipped with an osmotic absorption machine.
35. Describe the principle of operation of an osmotic absorption generator. What is its energy efficiency?
36. What is the difference between the hydrodynamic and magnetohydrodynamic technologies?
37. Which are the basic types of propellers used in hydrodynamic technologies?
38. What is the energy efficiency of the "ideal propeller" of Betz?
39. What is the functional relation between the aerohydrodynamic power and the fluid flow velocity?
40. How can one use hydrodynamic technology to recuperate heat from to a wind electric power station or from industrial technologies?

41. What are the requirements that a turbine generator should meet during pumping-accumulation recuperation?

42. What is the structure of the carriers of solar energy reaching the Earth's surface?

43. What the "photon number" of certain energy orbital is found from?

44. How can one change the characteristics of photon flux?

45. Which are the basic technologies used to change the characteristics of solar energy? What is the difference between them?

46. Which are the devices used in the realization of "reflective geometry"?

47. Which are the advantages of optical technology called "panel optical light conductor"?

48. What are the requirements of implementing regeneration and recuperation technologies?

5

Energy Transfer and Storage

5.1 From Spontaneous to Controlled Transfer and Storage of Energy (Trends in the Development of Energy Transfer Technologies)

Energy transfer and storage, in particular thermal energy, was a subject of physics until the end of the 19th century. At the beginning of the 20th century, however, it emerged as an independent branch of thermodynamic and power engineering. This was due to the remarkable results of L. Boltzmann, A. Einstein, P. Debai, N. Bohr, W. Heisenberg, I. Tam, Kaganov, D. Dzo, A. Majumder, and many other researchers in the field of fluid and quantum mechanics. Classical and modern studies cover the mechanisms of heat transfer in working bodies considering their three physical states—solid, liquid, and gaseous. Enormous experimental evidence has been also gathered as a result of many studies performed in research laboratories, universities, and companies. The results prove the capacity of various materials to receive, store, and transfer heat, as shown in Figure 5.1. It is seen that the material physical state is of crucial importance for heat conduction, and the coefficient of thermal conductivity λ, W/mK[*] varies as follows:

- For solids—$0.05 \leq \lambda \leq 500$ W/mK
- For fluids—$0.02 \leq \lambda \leq 5$ W/mK

The first important factor affecting material capacity to transfer and store heat is the operational temperature (T), and the subsequent empirical relation reads

$$\lambda = a_P T^2 + a_I T^{-1}.$$

Here, a_p and a_I are coefficients experimentally found for various materials. It is established that the thermal conductivity coefficient reaches a maximal value (λ_{max}) when the operational temperature is $T = (a_I/2a_P)$. The latter varies for different materials. It increases for materials which have high values of electrical conductivity and whose coefficient a_I is larger[†].

The property of bodies to transfer heat depends also on other macroscopic factors such as the Fermi energy level, the effective electric potential, the electromagnetic intensity

---

[*] The coefficients of thermal and electrical conductivity λ and Ω are directly proportional to each other pursuant to the Weidemann–Franz law—$\lambda = L_0 \cdot T \cdot \Omega$.

[†] This phenomenon is explained by the fact that the operational temperature T affects the width of the banned area and the emission of free electrons. The banned area according to modern physics is located between the valence band and the free area.

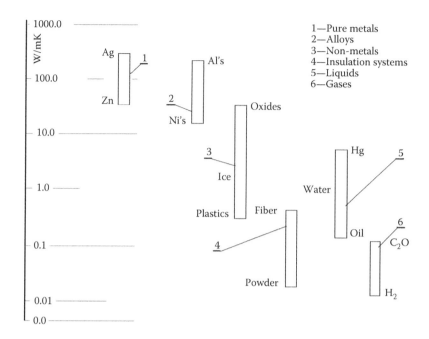

FIGURE 5.1
Range of variation of the conductivity coefficient of materials used in building structures, with respect to their aggregate state and chemical composition.

in the TDS control volume, enthalpy, and entropy, and the Gibbs, Helmholtz, Landau, Duhem, etc. energy potentials.

Another important material property is the capability to store heat. The so-called "kinetic thermal theory" predominated in the studies of heat transfer and in thermodynamics until the end of the 19th century, and its basis was the concept of material continuum. Later, however, it was found that solid bodies under low temperature (T < 100 K) cease to behave in accordance with the postulates of the kinetic theory. The most drastic departure from the predictions of this theory was the behavior of bodies having stored heat, assessed by the "specific heat capacity"—C_v J/m³ K.

It was found (Figure 5.2) that C_v depends on T in the temperature range $0 < T < 100$ K, and the dependence is nonlinear. For temperature close to absolute zero (T → 0 K), the specific heat capacity is zero. Then, C_v increases with the increase in temperature, and for $T/\theta_D \geq 2.1$ (T > 100 K),[*] it asymptotically approaches a constant value (C_v = 6.4 cal/gm–mol K). This value was experimentally found by Dulong and Petit in 1819—C_v = const ($C_v = 3N_A \cdot k_p/m = 3R \approx 6.4$ cal/gm–mol K) not depending on temperature.

The establishment of the nonlinear dependence of C_v on temperature generated the first problem in the understanding of energy storage in solids, since the kinetic gas theory was not able to explain this phenomenon.

A. Einstein (1905) and then P. Debye (1910) were the first researchers to do this. They assumed that discrete (quantum) processes took place in a discrete medium (consisting of

---

[*] θ_D—Debye temperature.

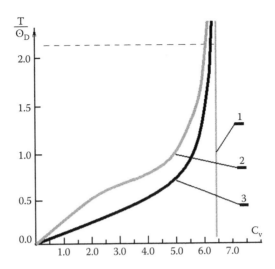

FIGURE 5.2
Variation of the specific heat capacity under low temperature: 1—approximation by means of DuLong/Petit law; 2—approximation by means of the Einstein model (quadratic approximation); and 3—approximation using the Debye model (cubic approximation in agreement with the experimental evidence for Pb, Ag, KCl, Zn, NaCl, Cu, Al, CaF$_2$, and C).

atoms and molecules).[*] Thus, they laid a new beginning of the theory of heat transfer,[†] that is, the study of heat transfer based on the concepts of quantum mechanics and statistical thermodynamics.

Without going into detail, note that energy transfer in solids takes place by means of two types of high frequency wave motion:

- *First type*: Energy transfer by means of longitudinal oscillations of atomic structures or by means of the so-called Einstein's waves (see Figure 1.2).

- *Second type*: Energy transfer by means of transversal waves (Debye's waves) where the lattice oscillates as a flexible membrane.

The behavior of the atomic lattice depends on the type of "impacts" penetrating the boundary of the TDS control surface (see Figure 5.3). If the amplitudes of the external disturbances are smaller than the interatomic distance "a," the lattice atoms oscillate *autonomously and independently*, that is, Einstein longitudinal waves propagate.

Otherwise, atom oscillations *correlate* with each other and the entire lattice behaves as a flexible membrane (i.e., Debye waves propagate). The ratio of the transferred energy is 1:2 in favor of the transversal oscillations under high temperature. This determines the nonlinear character of the solid body energy, which is analytically presented by a cubic approximation of the dependence $C_v(T)$—see Figure 5.2.

---

[*] Solids are built of periodically repeating atomic structures (about 70% of metals and alloys have simple cubic, face-centered cubic, body-centered cubic, or hexagonal crystal lattices).

[†] The entire solid composed of N atoms is modeled as a 3N multidegree harmonic oscillator, whose atoms (oscillators) vibrate autonomously with identical frequency v_E, that is, longitudinal harmonics with one and the same frequency propagate along the axes of a Cartesian coordinate system.

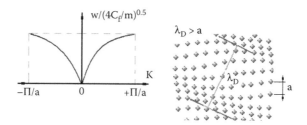

FIGURE 5.3
Relation between the angular frequency and the wave number.

Modern studies assume that solids' capacity to transfer energy depends on the frequency ν of vibration transfer. The renowned Russian physicist I.E. Tam[*] interpreted those vibrations as a result of vibration of pseudo-particles (quanta) called *phonons*.[†] His theory was universally recognized stating that *phonons are the basic energy carriers in solids.* They possess boson characteristics, and their behavior is described by the Bose-Einstein function.

According to modern scientific interpretation, $\bar{n}_{(\nu)}$ phonons may occupy a certain energy orbital (carrying frequency ν). The phonon number in different energy orbitals is determined by the local value of the Lagrange multiplier β_L ($\beta_L = (k_B \cdot T_L)^{-1}$), where k_B is the Boltzmann constant, or by the local temperature T_L. At low values of the Lagrange multiplier (under high temperature, respectively), phonons occupy high-energy orbitals and are carried by standing waves with small length.

When the value of the Lagrange multiplier increases (the local temperature $T_L\Downarrow$, respectively), phonons leave the high-energy orbitals and pass to low energy orbitals (long standing waves). Phonon motion between permitted energy orbitals is found according to the laws of quantum mechanics, depending on the energy generated and distributed in the TDS control volume. Those phonons (quasi-particles) are capable of interacting with the material particles—electrons (fermions), atoms, and molecules, and transferring energy to them. They are divided into two groups: thermal and acoustic. It is assumed that they are due to arbitrary energy fluctuations occurring at the boundary of the control surface. Just the thermal phonons with frequency of order (3–8) THz are responsible for energy transfer in bodies.

Four different microscopic level models are based on hypothetical mechanisms that were developed during the 1940s. These are

1. *Kaganov's Model or a two-step mechanism of interaction between phonons and electrons:*
   - *First step*; Heat transfer is initiated by the directed motion of electrons coming from the free area. Their motion (thermal) within the crystal lattice is characterized by a specific level of their internal energy $u_{El} = c_e T_e$, and respectively, by an internal kinetic potential (temperature of the electron gas T_e) and local Lagrange multiplier β_L.

---

[*] I. E. Tam was the 1938 Nobel Prize winner in physics.
[†] Phonons belong to the group of *bosons*, which are elementary particles with integer spin (other members of the boson group are photons, pions, gluons, and W- and Z-bosons).

- *Second step*: The kinetic energy of the moving electrons is transferred to the lattice atoms which start vibrating (i.e., generating phonons) with frequency ν depending on the temperature T_e of their equilibrium state.

2. *Model of Geier and Kumhasi—designed 10 years after Kaganov's model*: This model states that when the oscillating crystal lattice attains equilibrium (temperature T_1 and internal energy u_1), it starts emitting and scattering phonons creating heat flux with two components:

- Migrating phonons with sound velocity $\vec{c}^2$, the so-called elastic coalition. This flux is proportional to the gradient of the temperature T_1 of the solid crystal lattice T_1.
- Phonons forming plastic coalitions (similar to friction losses).

The energy is stored by the phonon system.

3. *A. Majumder's Model (1993)*: Heat exchange takes place by crystal lattice phonon emission. According to this model, the excited atomic lattice with temperature T_1 emits phonons similar to those emitted in compliance with the law of Stefan–Boltzmann, while the heat flux generated is described by the equation:

$$q = \frac{4\sigma}{3(L_0/1)+1}(T_1^4 - T_0^4),$$

Here, L_0 is width of the TDS control volume, while 1 is the phonon free path between two successive coalitions, and σ is the Stefan–Boltzmann constant (see Section 5.2.5).

4. *Model of lagging temperature gradient (Dimitrov, 2105)*: In the bodies under the influence of the environment (external energy fluxes or electromagnetic or gravitational fields) occurs directed movement of micro particles:

- Type fermions (electrons and ions)
- Quasi-particles (bosons type: photons or phonons by means of internal ionization or polarization of the atomic structures)

The movement of these particles in the transmission medium is hampered by the body's structural components (atomic lattices or molecular entities). This process is accompanied by transfer of an amount of movement from the quasi-particles (bosons) flow toward the lattice, whose internal energy (respectively temperature) is changed, proportional of the flux value. Thus, the change in body's temperature is caused by the directed movement of quasi-particles, delaying (lagging) in time relative to the passage of their flow.

The four models of heat transfer in solids contribute to the understanding of the physical character of transfer. They aid the design of a more precise description of this phenomenon and the derivation of more accurate engineering formulas.

The practical benefit of drawing a detailed physical picture of energy transfer in nature results in: (i) a capability to predict the TDS behavior and (ii) the development of new technologies for the regulation of heat transfer and conservation in engineering equipment (machines, buildings, and their components).

5.2 Control of Heat Transfer in Energy Equipment

There are formally three types of heat transfer at the macroscopic (engineering) level: thermal conductivity, convection, and radiation. We shall discuss in what follows their features and practical methods of their regulation by the introduction or avoidance of thermal resistance, such as volume or Fourier resistance, capacitive resistance (due to storage), aerodynamic resistance (due to convection), and reflection resistance (due to radiation). We shall start with quantities needed to assess the amount of energy transferred to a TDS, considering all practically established energy forms.

5.2.1 Vector Quantities of Heat Transfer and Gradient of the Temperature Field

5.2.1.1 Vector Quantities of Heat Transfer

Energy-related practice needs quantitative assessment of the motion of energy carriers, such as electrons, ions, or phonons in open TDS, whereas motion runs from areas with a higher free energy potential (indicated by higher temperature, pressure, and/or electric potential) to areas with a lower energy potential. That assessment specifies the intensity of energy transfer. It is the basis of all engineering algorithms of transfer organization or control in a TDS. The idea transfer modeling by vector presentation of its quantitative characteristics proves to be quite useful. Such vector characteristics of transfer are

- Energy

$$\text{Heat transferred, } \vec{H}, \text{J or electricity amount } \vec{\Xi}, \text{J.}$$

- Speed

Heat flux:

$$\vec{Q} = \lim_{\Delta 8 \to 0}\left(\frac{\Delta H}{\Delta \tau}\right) = \frac{d\overline{H}}{d\tau}, \text{J/s} = \text{W,}$$

or

Electric current:

$$\bar{I}_E = d\vec{\Xi}, \text{C/s} = \text{A.}$$

(5.1)

- Density

Specific heat flux:

$$\vec{q} = \lim_{\Delta A_0 \to 0} \frac{\Delta \overline{Q}}{\Delta A_0} = \frac{d\overline{Q}}{dA_0}, \text{W/m}^2,$$

or

Current density

$$\vec{\delta} = \frac{dI_E}{dA_0}, \text{A/m}^2$$

(5.2)

The integration of those equations yields the following relations:

- For heat amount:

$$\overline{H} = \int_{\tau_H}^{\tau_K} \overline{Q} d\tau \quad \left(\text{for electricity } \overline{\Xi}_E = \int_{\tau_H}^{\tau_K} \overline{I}_E d\tau \right) \tag{5.3}$$

- For heat flux:

$$\vec{Q} = \int_{A_0} \vec{q} dA_0 \quad \left(\text{for electricity } \vec{I}_E = \int_{A_0} \vec{\delta} dA_0 \right), \tag{5.4}$$

and respectively,

$$\overline{H} = \int_{\tau_H}^{\tau_K} \left(\int_{A_0} \vec{q} dA_0 \right) d\tau \quad \left(\text{for electricity } \overline{I}_E = \int_{\tau_H}^{\tau_K} \left(\int_{A_0} \vec{\delta} dA_0 \right) d\tau \right).$$

If transfer in a TDS runs in a 3D continuum of vectors $\overline{H}, \overline{Q}$ and $\vec{q}$, it seems useful to present them by their components in a Cartesian coordinate systems, for instance:

$$\vec{q} = \vec{q}_x \vec{i} + \vec{q}_y \vec{j} + \vec{q}_z \vec{k},$$

Yet, this presentation is not an object of the present book.[*]

5.2.1.2 Temperature Field

As already discussed in Section 1.1, TDS temperature being a basic state parameter holds information about the TDS internal kinetic energy and its importance for the objective assessment of the system general state. In most practical cases (in solids and incompressible fluids and at standard pressure), temperature distribution within the TDS control volume is specified by the boundary conditions. The set of values of T within the control volume is called the temperature field. It is formally described by a scalar function of the spatial coordinates of body points and time, having the following form in Cartesian coordinates:

$$T = f(x, y, z, \tau).$$

The following special cases of temperature field exist:

- *Steady field* (not depending on time): $T = f(x,y,z)$.
- *Plane field* (depending on two coordinates): $T = f(x,y)$.

---

[*] Curious readers may refer to the additional literature specified at the end of the present book.

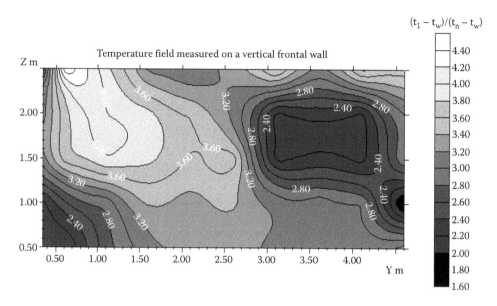

FIGURE 5.4
Isothermal lines of a two-dimensional temperature field measured on a flat wall.

- *One-dimensional (1D) field* (depending on one coordinate, along axis $\overrightarrow{Ox}$, for instance): $T = f(x)$.

The visualization of the temperature field is clearer by mapping the surfaces of constant temperature. These are the so-called isothermal surfaces shown in Figures 5.4 through 5.6. For so-called "singly connected fields" which prevail in thermotechnics, only one isothermal surface may pass through a point of space occupied by a TDS.

Hence, the isotherms are closed surfaces in the general case, or they end at the control volume surface. A normal vector $\bar{n}_T$ can be defined at each point which an isothermal surface passes through, and it is oriented to higher temperatures (Figure 5.5). For instance, the presentation of vector $\bar{n}$ in a Cartesian coordinate system is as follows:

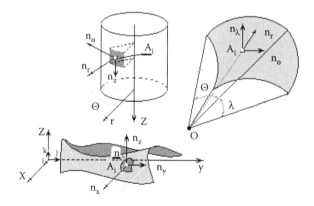

FIGURE 5.5
Isothermal surface of a 3D temperature field in a TDS. (The normal vector $\bar{n}$ directed opposite to the higher temperature and perpendicular to the isotherms is projected on the axes of three coordinate systems: Cartesian, cylindrical, and spherical.)

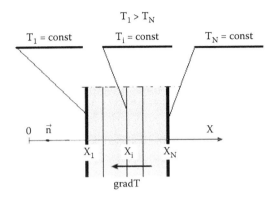

FIGURE 5.6
Isothermal lines of 1D temperature field having occurred within a flat wall. (The normal vector $\vec{n}$ is oriented opposite to the higher temperatures, and it is perpendicular to the isotherms.)

$$\vec{n}_T = n_{T_x}\vec{i} + n_{T_y}\vec{j} + n_{T_z}\vec{k},$$

where $n_{T_x}, n_{T_y}, n_{T_z}$ are projections of $\vec{n}$ on the axes Ox, Oy, and Oz, whose unit vectors are $\vec{i}, \vec{j}, \vec{k}$. The variation of temperature $T(x, y, z)$ along $\vec{n}$ is one of the most important characteristics of the temperature field, and it specifies the temperature gradient expressed by

$$\mathrm{grad}\,T = \lim_{\Delta n_T \to 0} \frac{\Delta T}{\Delta n_T} = \frac{\partial T}{\partial n_T} = \nabla T.$$

When the temperature field is 1D $T = f(x)$, the gradient can be written in the following simplified form:

$$\mathrm{grad}\,T = \frac{dT}{dx}\vec{i},$$

since the isotherms are straight lines perpendicular to the unit vector $\vec{i}$ (see Figure 5.6).

Otherwise, when T_1 is the highest temperature within the control volume (as is in Figure 5.6), the gradient will be negative with respect to the assumed orientation of the coordinate axis Ox.

5.2.2 Energy Transfer in Solids, Gases, and Liquids (Thermal Conductivity and Convection)

When the working body of a TDS is solid, the transferred energy is assessed by the total energy transfer law, giving the relation between the specific energy flux h and the gradient of free energy potential ψ^* in the solid body (Dimitrov, 2015). This law establishes that

$$\vec{h} \approx -\mathrm{grad}\,\psi. \tag{5.5}$$

---

* $\psi = (pv_0 - s_0 T + e_0 V)$ is the potential of a three functional power field where three types works: mechanical, thermal, and electrical are done, simultaneously.

The proportionality can be written in the vector equation form by

$$\vec{h} = -\bar{\Gamma} * \text{grad } \psi = -\bar{\Gamma} * \text{grad}(pv_0 - s_0 T + e_0 V), \tag{5.6}$$

or in matrix form by the product of two matrices:

$$h = -[\Gamma]\{\nabla\psi\}, \tag{5.7}$$

where $\bar{\Gamma}$ is a coefficient of proportionality, called general transfer coefficient.
The proportionality coefficient $\bar{\Gamma}$ reads as a diagonal matrix with rank 3:

$$[\Gamma] = \begin{bmatrix} \Gamma_T & 0 & 0 \\ 0 & \Gamma_M & 0 \\ 0 & 0 & \Gamma_E \end{bmatrix},$$

where Γ_T, $(ms)^{-1}$ is the coefficient of transfer of thermal energy, Γ_M, $(ms)^{-1}$ is the coefficient of transfer of mechanical energy, and Γ_E, $(ms)^{-1}$ is the coefficient of transfer of electrical energy. The matrix-column $\{\nabla\psi\}$ is with rank 3, whose components are the partial derivatives of the function ψ with respect to the ETS generalized coordinates—entropy s_0, volume v, and electric charge e_0 (*see* Chapter 1 and Section 1.1.3).

When the second and third components of "the general transfer coefficient $\bar{\Gamma}$" are zero, Equation 5.7 is transformed to Fourier's law[*]:

$$\vec{h}_T = \vec{q} = -\begin{bmatrix} \Gamma_T & 0 & 0 \\ 0 & 0 & 0 \\ 0 & 0 & 0 \end{bmatrix}\begin{Bmatrix} (\nabla\psi)_{s_0} \\ (\nabla\psi)_{v_0} \\ (\nabla\psi)_{e_0} \end{Bmatrix} = -\Gamma_T\nabla T = -\lambda \text{ grad } (T), \quad W/m^2. \tag{5.8}$$

For 1D temperature field shown in Figure 5.6, we have

$$\vec{q} = q_x\vec{i} = -\lambda\frac{dT}{dx}\vec{i}$$

or $\qquad\qquad\qquad\qquad\qquad\qquad\qquad\qquad$ (5.9)

$$q_x = -\lambda\frac{\Delta T}{\Delta x}\vec{i}$$

(since $dT/dy = 0 \rightarrow q_y = 0$ and $dT/dz = 0 \rightarrow q_z = 0$). Consider that for $x = x_1 \rightarrow T = T_1$ and for $x = x_N \rightarrow T = T_N$. Then, the integration of the differential equation within the boundaries of the TDS control volume yields

---

[†] If Γ_T and Γ_M are zero, Equation 5.6 is transformed to Ohm's law: $\vec{h}_E = \vec{\delta} = -\Gamma_E \cdot V = -\Omega \cdot \text{grad } (V)$, A/m^2.

$$q(x_N - x_1) = -\lambda(T_N - T_1)$$

or

$$q = \lambda \frac{T_1 - T_N}{x_N - x_1} = \lambda \frac{T_1 - T_N}{\delta_W} = \frac{T_1 - T_N}{(\delta_W / \lambda)} = \frac{T_1 - T_N}{R_W} = U_W(T_1 - T_N).$$

Here, $x_N - x_1 = \delta_W$ is "width" of the control volume or wall thickness while the ratio $\delta_W / \lambda = R_W$ is the wall thermal resistance. The reciprocal value of the thermal resistance is the wall thermal conductance called "U value": $U_W = 1/R_W$, W/m^2K.[*]

When the TDS physical medium is composed of several layers with different thicknesses, being fabricated from different materials (with different λ_i), the following equality is valid after the establishment of thermal equilibrium (see Figure 5.7):

$$q_1 = q_2 = q_3 = q. \tag{5.10}$$

Putting

$$q_i = \frac{(T_i + T_{i+1})}{R_i} = \frac{\lambda_i}{\delta_i}(T_i - T_{i+1})_i, \quad i = 1, 2, 3$$

in Equation 5.9, we obtain a system of three equations. Their summation yields:

$$q = \frac{T_1 - T_4}{\sum_1^3 \delta_i / \lambda_i}, \tag{5.11}$$

where T_1 and T_4 are wall surface temperatures and $R_{Wall} = \sum_1^n \delta_i / \lambda_i$ is the thermal resistance called volume (Fourier) resistance of a multilayer wall.[†]

When the working body is fluid (liquid or gas) and the boundaries of the control volume are close to each other ($\delta_{fl} = (x_1 - x_N) < \delta^*$), the adhesion intermolecular forces are those of friction (viscous forces). The fluid is immovable and the mechanism of heat transfer is

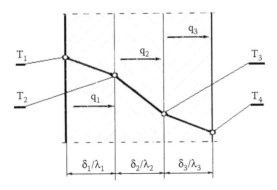

FIGURE 5.7
The specific heat fluxes become equal to each other after equilibrium is established.

---

[*] The values of U_W are normalized in structural design, and recommended (reference) values are used.
[†] In a cylindrical coordinate system, used to map an axisymmetric TDS, the thermal resistance is calculated via he formula $R_{CT} = \sum_1^n \ln(D_i / D_{i+1})/2\pi\lambda_i$ where Di is the diameter of the *i*th layer of the control surface.

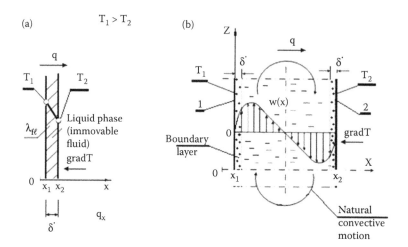

FIGURE 5.8
A TDS with liquid or gaseous working body: (a) 1D heat transfer and (b) simultaneous combined transfer.

similar to that taking place in the solid phase of the working fluid—Figure 5.8a. The heat flux is calculated using formulas (5.9), but assuming that $\lambda = \lambda_{fl}$ and $\delta_{fl} = x_2 - x_1$.

However, when the control volume expands to an extent where individual molecules move freely within the fluid (Figure 5.8b), *another very powerful mechanism* joins the already discussed conductivity. This is the mechanism of convective transfer.

Heat is transferred by fluid micro-volumes—they "extract" heat from the warmer boundary and supply it to the colder one. Circulation convective motion* occurs in the gap—if $T_1 > T_2$ motion is clockwise. Inversion of the gradient of the temperature field by 180° inverts the direction of convective motion. This is illustrated in Figure 5.8b. If temperature T_1 (at the left boundary of the TDS) is that of the HS ($T_1 > T_2$), the working fluid close to the wall is heated via thermal conduction through the boundary layer, and the specific heat flux q is proportional to gradT as stated by the Newton–Richman law describing heat transfer from a solid wall to a fluid. This can be written in the form

$$q_K = \frac{T_1 - T_{fl}}{\delta^*/\lambda_{fl}} = U_{out}(T_1 - T_{fl}), \tag{5.12}$$

where

- $U_{out} = \lambda_{fl}/\delta^*$ is a coefficient of convective heat transfer from the boundary wall 1. It is reciprocal to the thermal resistance of a boundary layer with thickness δ^* ($R_{fl} = \delta^*/\lambda_{fl}$)
- $gradT \approx (T_{fl} - T)/\delta^*$, $i = 1, 2\dots$
- $\delta^* = (z, v, T, \rho)$ is the thickness of displacement of the boundary layer depending on the velocity profile $w = f(z)$ ($\delta^* = 1.721\sqrt{(v \cdot z)/w}$ is the limit thickness of displacement of the boundary layer)

Convective motion can be generated by gravity force acting on a fluid whose density is different at the two boundary surfaces (1 and 2). It can also be externally generated by a pump or compressor, for instance.

The heated fluid becomes lighter ($\rho\downarrow$) and moves vertically upward. Thus, natural convection is initiated. An opposite process runs at the right boundary ($T_2 < T_1$), where heat is transferred from the warmed fluid to the cold wall. The fluid cools down, its density drops and it starts moving in an opposite direction. Thus, the fluid trajectory forms a circular contour (similar to a vortex deformed within the control boundaries).

The thermal conductance U_{out} depends on many factors (physical—$w,\lambda,\mu,\rho,c_\beta,\beta,T$ and geometrical—roughness Δ_{CT}, distance between walls 1 and 2 ($x_2 - x_1$), vertical coordinate z) and is found as

$$U_{out} = \frac{\lambda_{fl}}{\delta^*} \sqrt[3]{Gr \times Pr}, \quad \Rightarrow Gr \times Pr < 10^6.$$

Here

- $Gr = g \cdot \beta \cdot \delta^* \cdot \Delta T / v^2$—Grashoff number
- $Pr = \mu \cdot C_p \cdot g / \lambda_{fl}$—Prandtl number

and g—Earth's gravity, β—coefficient of volume expansion, μ—dynamic viscosity, C_p—specific heat capacity, $\upsilon = \mu/\rho$—kinematic viscosity.

If the working fluid is air (which is an excellent heat insulator), increase in the width of the control area ($x_2 - x_1$) yields decrease of the coefficient of thermal conductance U_{out}, while thermal resistance R_{fl} increases,* since $\delta^*_{r.c.}$ increases. Under standard conditions, both coefficients U_{in} and U_{out} amount to 10.0 and 25.0 W/m²K, respectively.

Thermal resistance within the boundary layer R_{fl} strongly depends on the density of the working fluid. For gases with large density (neon, argon, xenon, etc.), convective motion occurs under high temperature differences. This guarantees large values of the thermal resistance within gaps where inert gases are injected under normal temperature. The thermal resistance R_{fl} is large in gaps filled with rarefied gases or completely vacuumed (there the effect of convective transfer weakens and thermal conductivity is deteriorated due to the decreased number of carriers per TDS unit mass).

If a TDS is open and the fluid is assumed to be viscousless ($dp = 0$), the supplied heat δQ is entirely consumed to change the total internal energy. Then, the following form of the energy balance within the control volume holds, as found from Equation 1.34:

$$\delta Q = dH \tag{5.13}$$

(since $Vdp = 0$).

For 1D Cartesian control area $A_0 = 1$ m², the energy balance equation reads

$$-\frac{d}{dx}(q_x)dx\, d\tau = \left(\rho C_p \frac{dT}{d\tau} - q_v\right)dx\, d\tau, \tag{5.14}$$

since

- $dQ = \left[\left(q_x - q_x - (dq_x/dx) \cdot dx\right)\right] \cdot A_0 \cdot d\tau = -(d/dx)(q_x) \cdot d\tau \cdot dx$ is the heat supplied to the control volume

---

* For vertical air layers formed within a gap between vitreous structural elements (windows, window displays, and glazed doors) wider than 0.15 m, the thermal resistance R_{fl} drops since convective motion is activated ($W\uparrow$, $\delta\downarrow$).

- $dH = m^{\bullet}dh = m^{\bullet}C_p \cdot dT = (\rho dV_0)C_p \cdot dT = \rho C_p dT \cdot dx$ is the change of enthalpy (the total internal energy) of the working fluid within the control volume ($dV_0 = A_0 dx = 1 \cdot dx$)
- $q_v \cdot A_0 \cdot dx \cdot d\tau$ is the energy generated in/absorbed by the control volume
- $q_x = q_x^{Cond} + q_x^{Conv.} = -\lambda_{fl}(dT/dx) + C_p \rho w_x T$

A standard transformation of the balance equation yields

$$\frac{dT}{d\tau} = \frac{\lambda}{\rho C_p} \frac{d^2T}{dx^2} - w_x \frac{dT}{dx} + \frac{q_v}{\rho C_p}. \tag{5.15}$$

This is the basic differential equation describing the temperature variation within a 1D control volume of a TDS.

The following form is often used, too:

$$\frac{DT}{\partial \tau} = a\frac{\partial^2 T}{\partial x^2} + \frac{q_v}{\rho C_p}. \tag{5.16}$$

Here

- $DT/\partial\tau = (\partial T/\partial\tau) + w_x(\partial T/\partial x)$ is the substantial change of temperature T at a TDS point. It is due to the local unsteadiness $\partial T/\partial\tau$ and to the convective temperature change $w_x(\partial T/\partial x)$ introduced along the axis x; $a = \lambda_{fl}/\rho C_p$—thermal conductivity number.
- $q_v/\rho C_p$—in-volume generation term.

For steady convective flows, without generation of energy within the TDS control volume, Equation 5.16, a is simplified in the form

$$w_x \frac{\partial T}{\partial x} - a\frac{\partial^2 T}{\partial x^2} = 0, \tag{5.17}$$

and if w_x = const (for $dA_0 = 0$) one gets the following differential equation for the temperature variation at a TDS point:

$$\frac{a}{w_x}\ddot{T} - \dot{T} = 0. \tag{5.18}$$

It can be easily integrated if the boundary conditions at the control surfaces 1 and 2 are known. For brevity, we differentiated temperature with respect to the Cartesian coordinate x, and the first and second-order derivatives have forms $\dot{T} = \partial T/\partial x$ and $\ddot{T} = \partial^2 T/\partial x^2$, respectively.

The simplified equation of convective heat transfer in a TDS shows that despite the boundary conditions, the temperature distribution will also depend on the physical properties of the working body (λ_{fl}, ρ, C_p) presented by the thermal conductivity number $a = \lambda_{fl}/\rho C_p$ and the speed of body motion. Equation 5.18 is used to predict the temperature distribution and the temperature gradient in 1D open TDS. More complex physical and mathematical models are used in two-dimensional (2D) and 3D cases of thermal conductivity, which are out of the range of the present book. Curious readers may refer to the literature specified in the References.

5.2.3 Characteristics of Heat Transfer through Compact Inhomogeneous Multilayer Walls

The above physical picture of heat transfer via conductivity and convection within a TDS control volume turns out to be idealized and incomplete to describe transfer through real compact structural elements: external walls, floor, and ceiling plates.

The real picture differs from the idealized one due to two main reasons:

- Technological (see Figure 5.9)
- Architectural-construction flaws

All deviations from and violations of the structural requirements belong to the first group. The second group of violations is (i) inclusion in the building envelope of materials

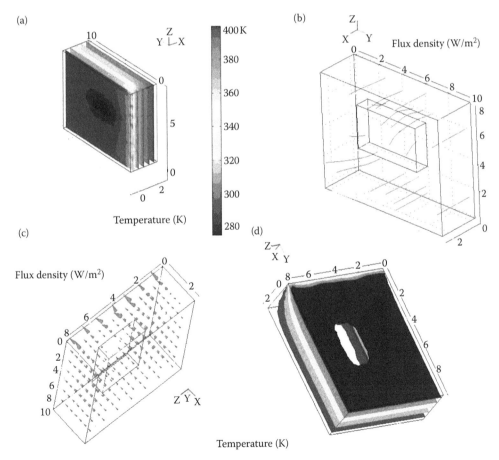

FIGURE 5.9
Structural inhomogeneities of a solid wall due to technological errors: (a) Temperature distribution; (b) energy flux lines; (c) directions of energy flux; and (d) thermal bridges in the wall. (Simulation of the heat transfer in an inhomogeneous physical environment employing the Comsol software.)

whose thermal conductivity essentially differs from that of standard structural materials, for instance steel fiber reinforced columns, beams, etc. (see Figure 5.10) and (ii) architectural design of buildings with broken up geometry—involving jetties, angles, counterangles, etc. To predict the amount of heat transferred through such complicated and inhomogeneous structures, thermal bridge models are often used. These are structures which provide additional heat transfer through the control surfaces *exceeding* transfer in 1D homogeneous medium described in Section 5.2.2.

A principle of superposition of heat fluxes (those running through the homogeneous foundation and through the bridges) is adopted for real structural elements

$$\vec{H}_{inhom} = \vec{H}_{hom} + \vec{H}_{bridge}. \tag{5.19}$$

Here

- $\vec{H}_{inhom}$ J—heat transferred through a real wall
- $\vec{H}_{hom}$ J—heat transferred through a homogeneous wall
- $\vec{H}_{bridge}$ J—heat transferred through thermal bridges

The heat flux $\vec{Q}$ (the rate of heat transfer) through a structural element with surface area A_0 reads

$$\vec{Q}_{inhom} = \vec{Q}_{hom} + \vec{Q}_{bridge},$$

where

- $\vec{Q}_{hom} = (\Sigma_1^H \lambda_i/\delta_i + U_{out} + U_{in})A_0(T_i - T_e)$—heat flux through an element without a thermal bridge
- $\vec{Q}_{bridge} = (\Sigma_{j=1}^k l_j\,\psi_j + \Sigma_1^m \chi_k)(T_i - T_e)$—heat flux through thermal bridges
- $\Sigma_{j=1}^k l_j \cdot \psi_j$—additional heat flux through linear thermal bridges l_j,m, intensity ψ_j, W/mK
- $\Sigma_k \chi_k$—additional heat flux through point thermal bridges with intensity χ, W/K

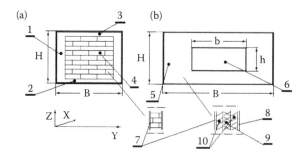

FIGURE 5.10
Structural inhomogeneities due to the use of structural elements with different thermal conductivity. (a) Fasade brick wall and (b) panel wall: 1—steel fiber reinforced column ($\lambda = 1.63$ W/mK); 2—steel fiber reinforced plate ($\lambda = 1.45$ W/mK); 3—steel fiber reinforced beam ($\lambda = 1.63$ W/mK); 4—brick wall ($\lambda = 0.52$ W/mK); 5—steel fiber reinforced panel, sandwich type ($U_{cm} = 0.4$ W/m² K); 6—glazed element ($U_w = 2.6$ W/m² K window with wooden frame); 7—lime-cement putty ($\lambda = 0.81$ W/mK); 8—internal plaster ($\lambda = 0.87$ W/mK); 9—foam-polystyrene ($\lambda = 0.03$ W/mK); and 10—steel fiber reinforced concrete ($\lambda = 1.63$ W/mK).

The temperature difference $(T_i - T_e)$ is the operational link between the "cold" and the "hot" source of a TDS.

5.2.4 The Role of Heat Storage in Compact Bodies: Capacitive Thermal Resistance

All TDS tend to store such amount of heat Δq_{In}, so that entropy would attain its maximum under specific operational conditions, and the TDS would attain equilibrium (second principle of thermodynamics). Thus, the attained level of TDS entropy s_i is the sum of the initial entropy s_0 plus the stored energy, that is, the energy charge[*]

$$s_i = s_0 + \frac{\Delta q_{In}}{T_1} = s_0 + \frac{\Delta h_{0-i} + r_{PhE}}{T_1}, \qquad (5.20)$$

where

- Δh_{0-i}—increase of the total internal energy reforming (entropy) of the working body with accumulating mass m, kg and specific heat capacity C_p, J/kg K
- T_1—operational temperature of the hot body, K
- r_{PhE}—heat of phase transition of the working body, J/kg

The possibility of changing the *current value of entropy* s_i and the subsequent use of the stored energy is a serious challenge to designers of energy-generation systems, since the TDS operates under variable temperature difference $(T_1 - T_2)$. The change of entropy current value s_i should be used as an energy buffer.

When there is excess of energy in the "hot" energy source and it exceeds the energy used in a TDS, the increase of entropy current value s_i yields storage of the unneeded energy into the TDS for further use. Then, when the energy demand exceeds energy supply from the hot energy source, the energy needed is supplied at the expense of entropy s_i, stored in the energy buffer of the heat accumulator. Considering Equation 5.20 for entropy variation, we can classify the heat storages as follows:

- Enthalpy storages
- Latent storages (phase-exchanging—PEA or chemical)

Enthalpy storages are traditionally used in power engineering installations operating with renewable sources (RES). Solar installations (active and passive), wind electo-generators, and machines consuming "night" and "season" energy need a heat storage due to the stochastic nature of the external energy. The latter should be used if available (solar energy should be used in the midday, and wind power should be used depending on the installation latitude and longitude: in the morning, at dusk, or at night). If there is no current energy demand, energy should be stored for further use.

The storage capacity is selected varying the accumulating mass m_{TDS} and the operational temperature difference within a wide range, but without reaching the temperature

---

[*] It is a priori assumed that the stored heat is spent to change the total internal energy and the aggregate state of the working body.

that would yield a change of phase in the working body or activate chemical changes (reforming). The working medium should have a high coefficient of thermal conductivity and large ratio C_v/V (heat capacity per unit volume) to guarantee thermal homogeneity within its entire volume. The working medium material should be accessible and not expensive. The following materials are the appropriate ones for this purpose: fire resistant concrete (C_v/V = 1963 kJ/m³ K), fireclay (C_v/V = 2900 kJ/m³ K C_v/V = 2900 kJ/m³K), olivine (C_v/V = 2730 kJ/m³ K), water (C_v/V = 4186 kJ/m³ K), magnezite (C_v/V = 3333 kJ/m³ K), cast iron (C_v/V = 3915 kJ/m³ K), and oil (C_v/V = 1630 kJ/m³ K).

Latent storages are divided into phase-change and chemically latent ones. *Phase-changing latent storages* (PELA) operate under constant temperature, which is a good prerequisite for profitable energy saving (without additional losses). There are a number of chemical compounds which meet technological requirements similar to those for enthalpy storages, but practically, $NaNO_3$ is the most often used compound.

Compounds that pass from solid to gaseous aggregate state in the temperature range $18°C < T < 28°C$ are of special practical interest. Also, this range meets the standard requirements for thermal comfort in inhabited areas.

Chemically latent storages use as working medium compounds which are capable of participating in reversible chemical reactions. They operate in a two-phase cycle: heat storage (endothermal phase) and heat release (exothermal phase).

In the *endothermal phase*, the storage is charged with heat, accumulating energy from the hot energy source. During the *exothermal phase*, it releases heat to the CS. The chemical storages operate within a closed thermodynamic cycle, and they have significantly higher thermal capacity than the enthalpy ones and the PCA. At the same time, energy conservation proceeds indefinitely under temperature close to the ambient one.

Storage effects in building envelopes are of special practical interest for the building thermal design. Envelopes are fabricated from materials with significant thermal capacity—concrete, steel fiber reinforced concrete (C_v/V = 2100 kJ/m³ K), stone (C_v/V = 2500 kJ/m³ K), brick masonry (C_v/V = 1600 kJ/m³ K), and plasters (C_v/V = 1400 kJ/m³ K). Knowing the mechanisms of heat storage in the external layers of structural elements is essential for the improvement of their energy-saving function.

Treat buildings and their elements from a *thermodynamic point of view*. Consider them as being TDS with a control volume equal to the building volume and with a control surface identical with the external surface of the building envelope. Then, it is found that under steady conditions (T = const), heat capacity of the working body (solid structural materials in this case) per unit area of the control surface a_0, reduces the heat flux passing through the building volume without time shift.

Assume change of the temperature boundary conditions T_e at the surface of a compact wall element with zero heat capacity. Then, the subsequent temperature response T_n is expressed by the decrease of the amplitude of T_n with respect to T_e.

Change of T_e and T_n takes place *simultaneously* or *with phase delay*. This was clarified in detail in Section 5.2.3. If the ambient temperature T_e varies in time, thermal resistance is actuated based on the heat capacity of the structural materials.[*]

---

[*] The technique of using the thermal capacity is successfully applied in passive solar architecture, having emerged in the 1980s, but originating from the ancient civilizations of Mesopotamia and Assyria. The energy system of passive solar buildings is integrated with the structural components. Renewable energy sources (the Sun, wind, Earth energy, etc.) are used to provide thermal comfort.

Figure 5.11 shows the reaction of the compact elements of a building envelope to harmonic change of the temperature of their external surface. We assume for convenience that the external temperature varies following a sine law[*]:

$$T_e = A * \sin\left(\frac{2\pi}{\tau_0} * \tau\right).$$

Here

- A, °C is the amplitude of temperature oscillation (see the example in Figure 2.64, A = 15°C = const).
- τ^0, h—period of temperature oscillation (plotted by three harmonics τ^0 = 24 h, 12 h, 6 h). The period τ^0 = 24 h is chosen to simulate the mean round-the-clock course of the temperature of the surroundings—see Figure 5.11).

Divide virtually the wall element into layers with identical thickness $\Delta\delta_{CT} = \delta_{CT}/(n-1)$. *Assume initially* that the entire wall has one and the same temperature and that the surface temperatures of the internal layers are identical, that is, $T_i = T_{\tau=0,h} = 15°C$.

If temperature T_e varies harmonically, the thermal-capacitive element will operate in two stages:

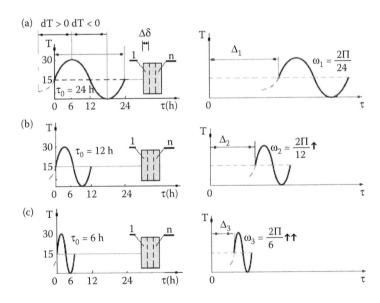

FIGURE 5.11
Effect of the heat capacity (ρC_v) on the temperature of the "internal surface" (T_n) for ρC_v = const (ρC_v = 3000 kJ/K). (The parameter of variation of temperature T_e is the period of temperature oscillation $\tau_0 \rightarrow \tau_0$ [24 h, 12 h and 6 h].)

---

[*] In reality, it varies following a quasi-periodical law, and it can be presented as a sum of harmonic oscillations with different frequencies corresponding to the round-the-clock and annual cycles, and as a sum of other overlapping random impacts.

- Transfer with the increase in T_e (within the range $-0.5\pi \sim 0.5\pi$)
- Transfer with the decrease in T_e (within the range $0.5\pi \sim 0.5\pi$)

During the first stage the temperature gradient will be positive (unidirectional with normal $\bar{n}$) while vector $\vec{q}$ will be negative (see Figure 5.11a). Heat flux $\vec{q}$ enters the element first layer and heats it (if the initial temperature is $T_1 = 15°C$, it reaches T_2 after heating): $T_2 = T_1 + q/C_v$.

The technological time of attaining T_2 depends on

- Heat capacity of the first layer ($C_v^* \Delta \delta_{CT} A_0 \rho$)
- Magnitude of the specific flux $\vec{q}$

The heating time is measured in picoseconds, but heat transfer to the wall slow down. When the first layer heats up to T_2 due to the positive temperature gradient, the second layer heats from T_1 to T_2, and the first layer increases its internal energy and temperature to T_3, since $T_e\uparrow$, proceeding to store heat. Thus, the internal energy increases from layer to layer retarding heat transfer. The retardation within the entire thermal-capacitive element is the sum of retardations in each layer. Although that significant total retardation may be established (Δ_1—Figure 5.11a), the temperature decrement T_n is small ($T_n \approx T_e$), since the thermal conductivity coefficient of structural materials is high—$\lambda = 1.63$ W/mK for steel fiber reinforced concrete and $\lambda = 3.5$ W/mK for natural stone.

Hence, the thermal-capacitive element is charged with heat as a result. *Summarizing* the above considerations, the thermal-capacitive element only retards heat transfer, without significantly affecting temperature decrease. In fact, it impedes heat transfer and plays the role of a specific thermal resistance similar to an electric capacitor.

During the second stage, the thermal-capacitive element discharges (the stored heat is spontaneously released to the surroundings or transferred to the interior). This process starts at the moment when the external temperature T_e, having passed through its maximum, starts decreasing.

Since at this moment the temperature of the first layer $T_1 \rightarrow T_e^{max}$, and T_e starts decreasing, the temperature gradient inverts its direction becoming negative with respect to $\bar{n}$. The specific heat flux $\vec{q}$ also inverts its direction—it is directed from the first layer to the surroundings, since the layer is warmer. This yields decrease in layer temperature and inversion of the gradient and the direction of the heat transfer from the second to the first element (Figure 5.11a). Thus, the direction of heat transfer is inverted layer by layer within the entire wall, while the internal temperature T_e decreases to T_e^{min}. Since temperature T_e varies after a harmonic law with period $\tau_0 = 24$ h, it is obvious that the thermal-capacitive element will operate within a cycle comprising *two processes*: charge and discharge. An extremely important question arises here, that is, whether discharge will take place unidirectionally and unilaterally (directed to the surroundings, only) or *bilaterally* (directed to the interior, too).

For a fixed element (with overall dimensions, mass, specific capacity C_v) and technologically found time of charge (τ_{Charge}), the key to the answer lies in the periodicity τ_0 of the external temperature harmonic (Figure 5.11b and c) illustrate the capability of other temperature harmonics to "charge" that same thermal-capacitive resistance. The example considered assumes that their periodicity is $\tau_0 = 12$ h and 6 h, respectively. Plots show that the time of charging the thermal-capacitive element (semi-period) is 200% and 700% smaller than that of the reference case $\tau_0 = 24$ h (Figure 5.11a).

Hence, it is obvious that if there are elements of the building envelope with small heat capacity, the temperature wave, although in a short time, can pass through the entire control volume reaching the last nth layer (see Figure 5.11b and c).

Assume that it has penetrated the volume "reaching" the last layer (although with a certain delay ($\Delta_3 < \Delta_2 < \Delta_1$)). Then, the layer heats/cools down. Since the last layer is an *"external boundary"* of an inhabited building volume, it interacts with the "interior," radiates/absorbs long-wave energy and heats/cools down the air boundary layer via thermal conduction. Then, the total thermal-capacitive element will operate bilaterally, receiving and releasing heat through its two boundaries (through the first and the last layer). Assume now that the heat capacity of the element is significant and the necessary charging time τ_{Charge} exceeds the semi periods ($\tau_{0/2}$) of the temperature harmonics. Then, the internal "temperature" wave would not be able to reach the nth layer and the external energy would not change the element thermal status. Hence, the element would operate unilaterally (it will be charged and discharged through its first layer, only). This circumstance is used to design compact elements of a building envelope, type "sandwich"—see Figure 5.13c.

It is obvious that the capacitive resistance will be inversely proportional to the specific capacity C_v of the working body and to the angular frequency $\omega_0 (2\pi/\tau_0)$ of the temperature oscillation

$$R_C = \frac{1}{(C_v \omega_0)}. \tag{5.21}$$

The mechanism thus outlined shows that a thermal-capacitive element (similar to an electric capacitor) functions as a thermal filter during nonsteady processes. It displays small resistance to long-wave and large resistance to short-wave thermal oscillations. Hence, compact wall elements with significant heat capacity:

- Transmit temperature fluctuations with large period (long-wave ones)
- Absorb (return/keep) fluctuations with a small period (short-wave ones)

The variable character of the specific heat flux q_{ω_0} passing through the compact element possessing a specific frequency form (ω_0), is described via a harmonic function

$$q_{\omega_0} = q_0 \sin(\omega_0 \tau), \tag{5.22}$$

and the respective change of the external layer temperature displays phase delay

$$T_i = T_0 \sin(\omega_0 \tau - \Delta_n). \tag{5.23}$$

Since the real oscillations of temperature T_e are composed of a number of frequency forms $(\omega_0)_i^*$ the energy transfer through the components of the envelope are interpreted as a sum of two components: $q = \bar{q} + \tilde{q}$.

---

* The variable temperatures in the solids walls are caused by variable phonons flus, according the model of lagging temperature gradient (Dimitrov, 2105).

Here

- $\bar{q} = 1.65A_0\,0.608(T_m - T_i)/0.865 + \delta_{cr}/\lambda$—specific heat flux, constant in time, whose value depends on temperatures T_m and T_i and on the volumetric thermal expansion of the envelope element—δ_W/λ

- $\tilde{q} = 1.65A_0\sum_{n=1}^{\infty}\Delta_n T_n * \cos(15n\tau - a_n - \phi_n)$—heat flux variable in time $\tau(12\,h \Rightarrow \tau = 0)$, presented as a superposition of n ($1 < n < \infty$) harmonics; Δ_n and t_n are attenuation coefficients and phase delay coefficients of the nth harmonic[*]

The final effect of the thermal capacity of a working medium (a compact structural element of the building envelope in our case) consists in *retardation and suppression* of the penetration of external temperature oscillations into the interior.

The heat capacity of envelope compact elements should be appropriately calculated, so that the instantaneous temperature changes outside or inside the building would not yield "overconsumption" of primary or supplied energy. Thus, the energy efficiency of the building envelope and that of the entire building could be improved. This circumstance *is not accounted for* in actual normative documents and codes, while it has a very good potential to save energy and is an important architectural-engineering instrument. Besides, disregard of the heat capacity of the building envelope implies an extremely difficult role for systems controlling the building thermal comfort. This is so, since the system automation will react to the "highest-frequency" temperature harmonics. To compensate for them, the temperature should be continuously varied and the supply to the inhabited areas with conditioned air should be appropriately regulated. Yet, this will compromise the thermal comfort and the operation of the respective building systems. Inhabitants of such "sick" buildings will tire easily and/or fall ill.

Hence, the effect of the thermal capacity of a building envelopes is often made used not on the basis of engineering calculations and estimations, but intuitively. We shall discuss what follows some practically applied techniques of utilizing intuition. Yet, we shall first analyze the third mechanism of heat transfer, that is, *thermal radiation*.

5.2.5 Radiative Heat Transfer: Reflective Heat Resistance

The mechanisms of heat transfer described so far are typical for solid and liquid bodies functioning as a TDS medium. However, the basic form of energy transfer in fluids or a vacuum is radiation. As already discussed, the energy carriers—boson quasi-particles, move as standing waves (occupy energy orbitals) emitted by all material bodies whose temperature exceeds the environment mean temperature.

The basic energy carriers in a vacuum are photons, which are of the boson type. They are the basic component of solar radiation and as discussed in Section 4.3.3, they are quasi-particles with wave properties. Under appropriate conditions, they display their corpuscle properties, too—for instance, a capability to exercise pressure.[†]

Bosons and photons in particular, which have the same energy, occupy one and the same orbital (one and the same energy saddle, metaphorically speaking). The number

---

[*] $\Delta_n = (1.41(F^2 + G^2)^{-1})$ and $\phi_n = \text{arctg}(F - G/F + G)$, $F = (r(q_0/q_c) + \pi_1 c_C(q_0/q_c))$, $G = (S_0(q_0/q_c) + \pi_1 S_C(q_0/q_c))$, and $S_{0,C} \rightarrow f(\rho, \bar{\lambda}u\,a)$.

[†] During nuclear reactions (see Figure 2.46) running within the Sun, alpha-particles, positrons, and neutrino are emitted besides photons. However, they do not reach the Earth.

of bosons in the orbital can be changed, depending on the value of Lagrange multiplier.[*] Arranged in a set, the bosons form standing electromagnetic waves with different length.

Although according to quantum theory, the photon energy charge $\hbar$ ($\hbar = h/2\pi = 1.05 \times 10^{-34}$ J s) is constant, and its impulse $\bar{p}$ depends on the length of the carrying wave (its energy orbital), and it is calculated as a product of $\hbar$ and k

$$\bar{p} = \hbar * k = \frac{h}{\lambda}.$$

Here, the orbital wave number k is specified by λ via the expression: $k = 2\pi/\lambda$.

The value of the photon impulse depends on the amount of carried energy. It is calculated using the formula $\langle e_{Ph} \rangle = \hbar \times \omega_0$ which shows that the photon kinetic energy also depends on the occupied orbital (here ω_0 is the angular frequency of the energy orbital occupied by the photon under consideration). Hence, the energy carried by a photon will depend on the occupied orbital, only, that is, on the respective standing wave (on the length λ of the carrying wave or on the wave number k, where $\lambda = 2\pi c/\omega_0$ and $c = 2 \times 10^8$ m/s).

A different number of photons $n(\omega_0)$, however, may occupy a certain energy orbital (specified by ω_0, k, or λ). This depends on the state of the photon gas characterized by the Lagrange multiplier β_{Ph}, respectively, the value of photon gas temperature T_{Ph} [$\beta_{Ph} = (k_B \cdot T_{Ph})^{-1}$, $k_B = 13.73 \times 10^{-27}$ kJ/K].

When T_{Ph} decreases (respectively, $\beta_{Ph}\uparrow$), photons leave the energy orbital and "settle" on an orbital with smaller wave number k. When T_{Ph} increases photons shift to orbitals with a larger wave number (to shorter waves). Photons "jump" from one orbital to another and stay on an orbital, which corresponds to the instant value of the Lagrange multiplier. Thus, they display their quantum character.

The energy of $\bar{n}(\omega_0)$ photons of a certain orbital with angular frequency ω_0, is a sum of the energies of each photon, and is calculated as

$$\langle E_{(Ph)} \rangle_{\omega_0} = \sum_{\substack{\omega_0 = const \\ n(\omega_0)}} \langle e_{ph} \rangle = \bar{n}(\omega_0)\langle e_{ph} \rangle = \bar{n}(\omega_0) \times \hbar \cdot \omega_0.$$

Since photons, as well as other bosons with zero spin, do not interact with each other if they belong to different orbitals, the distribution function $\bar{n}(\omega_0)$ has a form specified by Bose

$$\bar{n}(\omega_0) = (e^{(\beta\hbar\omega_0)} - 1)^{-1}.$$

Then, the orbital energy will amount to

$$\langle E_{(Ph)} \rangle_{\omega_0} = \frac{1}{8} \sum_0^{\sim} 2\frac{\hbar\omega_0}{e^{\beta\hbar\omega_0} - 1} = \frac{\hbar}{\pi^2 \cdot c^3} \cdot \omega_0^3 (e^{\hbar\omega_0/k_BT} - 1)^{-1}. \tag{5.24}$$

---

[*] The bosons migrate to short length waves when the system is energized and they move to the long wave in the opposite case.

Here c is photon velocity in vacuum. The expression found for $\langle E_{Ph} \rangle_{\omega_0} = f(\omega_0)$ is known as the Planck integral.

It is a fundamental quantity of all branches of physics, since it specifies orbital energy distribution considering orbital wave characteristics—λ, k, or ω_0, as well as the energy charges of the photon gas (characterized by the Lagrange multiplier β).

It seems noteworthy that the function $\langle E_{Ph} \rangle_{\omega_0} = f(\omega_0)$ has an extremum at $(\omega_0)_{max} = 2.82/(\beta \times \hbar)$. In heat transfer, the Planck integral was calculated by Stefan and Boltzmann, and the solution found is known as the Stefan–Boltzmann law. It illustrates the relationship between the integral amount of energy radiated by a perfectly black body along all orbitals and the radiation temperature (i.e., that of the photon gas $T_{Ph} = (k_B * \beta)^{-1}$).

According to the Stefan–Boltzmann law, if a perfectly black body is heated to temperature T, K, the energy radiated within the whole frequency range (i.e., from all energy orbitals) amounts to

$$E_{Abs} = C_S \left(\frac{T}{100} \right)^4 = 5.77 \left(\frac{T_I}{100} \right)^4, W/m^2. \tag{5.25}$$

For real bodies, this formula is completed by the so-called coefficient of the emissivity ε ($\varepsilon = C/C_S$, and C, W/m^2K^4 is a radiation coefficient for real bodies, which is experimentally found), and

$$E_{Grey} = \varepsilon * E_{Abs} = 5.77 \, \varepsilon * \left(\frac{T}{100} \right)^4. \tag{5.26}$$

Together with the estimation of energy radiation, it is worth estimating the absorption of photon energy by neighboring bodies and the possibilities to control this process. These are important issues from a technical point of view, and related problems can be solved employing the methods of analytical geometry (see Figure 5.12). For this purpose, we assume that energy E_0, radiated by "hot" bodies, propagates linearly (ray-wise) within all eight quadrants of the Cartesian coordinate system.

The energy E_φ carried by the total photon flux E_0 along the normal is $E_\varphi = E_0/4\pi$, where the energy radiated in space in only one direction, specified by a spatial angle φ, is found as

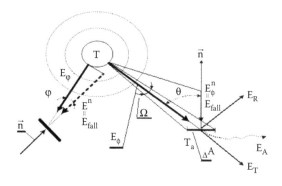

FIGURE 5.12
Radiation on surrounding surfaces.

$$E_\varphi^{Sh} = E_0 \cdot \frac{\varphi}{4\pi},$$

and in the plane case it reads

$$E_\varphi^F = E_0 \cdot \frac{\varphi}{2\pi}$$

(φ is the plane angle in Figure 5.12).

The amount of radiation E_{fall} having landed on a body arbitrarily located in space can be accessed via the so-called cosine law of Lambert (see Figure 5.12)

$$E_{fall} = E_\phi \cdot \cos\theta = E_\phi^\pi, \tag{5.27}$$

proving that only part of the radiation E_φ, lands on the irradiated surface, whose normal is $\vec{n}$.

The energy flux meeting the receiving surface is absorbed by, reflected from or passes through the receiving body, and the following balance equation holds:

$$E_{fall} = E_A + E_R + E_T. \tag{5.28}$$

Division of both sides by E_{fall} yields

$$A + R + T = 1. \tag{5.29}$$

Here A, R, and T are coefficients of absorption, reflection, and transmissivity of real bodies. These parameters are found experimentally. The values of the coefficients A, R, and T depend[*] on the chemical composition of the bodies, on the quality of the absorbing surface—color, roughness, and on the radiation wavelength.

Various materials have "selective" spectral properties. For instance, liquids and glass are not transparent materials with respect to the long-wave photons (including the infrared ones). Vitreous elements, excluding those fabricated from "quartz" glass, absorb also the ultraviolet (UV) photons. Depending on the color, colored surfaces reflect part of the radiation (blue reflects the short wave radiation; red—the long wave radiation; and white—the visible light [360–860 nm]). Hence, three radiation models are used in heat engineering at the macro-level: perfectly black body (A = 1, R and T = 0), perfectly transparent body (A = R = 0; T = 1), and perfectly white body (R = 1, A and T = 0).[†] The following relationship between the absorbed energy (e_a) and the energy falling on neighboring bodies along normals ($\vec{n}$)($E_{fall} = E_\phi^n$) exists for most bodies:

$$E_A = E_{fall}(1 - R - P).$$

---

[*] The values of the coefficients are published in engineering reference books.
[†] Idealized model of reflective resistance.

In fact, the energy exchanged between two bodies will depend on the temperature difference $(T_1^4 - T_2^4)$. If a TDS consists of two bodies, only, heat exchanged between them via radiation is expressed as[*]

$$Q_{1-2} = Q_R = A_{eff} * q_{1-2} = 5,77 \; \varepsilon * \left[\left(\frac{T_1}{100}\right)^4 - \left(\frac{T_2}{100}\right)^4\right] * A_{eff}$$

$$= U_R * A_{eff} * (T_1 - T_2) \qquad (5.30)$$

or

$$Q_R = U_R * A_{eff} * (T_1 - T_2).$$

Here T_1 and T_2 are body temperatures, ε is the degree of body grayness, and U_R is called coefficient of radiation heat release

$$U_R = 5,77 \; \varepsilon * \frac{(T_1^4 - T_2^4) \cdot 10^{-8}}{T_1 - T_2}, \; W/m^2 K. \qquad (5.31)$$

It is similar to the coefficient of convective heat release U_{Out}, U_{In}, and to the coefficient of thermal conductance $U_{thermal\;conductance} = \sum_1^n \lambda_i / \delta_i$ for multilayer compact walls. The reciprocal value of U_R is the reflective resistance or the resistance against radiative heat transfer.

Yet, since the process of heat transfer is complex, the engineering practice uses the sum of all coefficients, called *total coefficient of heat transfer* or *overall heat transfer coefficient*

$$U = U_R + \sum_1^n \left(\frac{\lambda_i}{\delta_i}\right) + U_{In} + U_{Out}. \qquad (5.32)$$

The effective surface area of heat exchange A_{eff} in Equation 5.30 is a complex quantity which, except for the body's geometrical surface, accounts also for the bodies' mutual disposition (curious readers may refer to the additional literature at the end of the book for more details).

5.2.6 Control of the Heat Transfer through TDS Control Surfaces (Arrangement of the Thermal Resistances within the Structure of Compact Wall Elements)

Since the time of its first successful steps and the great discoveries of the 19th century, heat transfer became an independent science and matured in the 20th century together with the electrotechnical and chemical sciences. This was the period when all modern power engineering technologies were designed, and essential problems of "How to do it" were solved. Material science, the physics of semiconductors, electronics and automation and computer sciences marked significant progress during the last 30 years. Thanks to this progress and low prices, manufacture of a wide variety of components of energy equipment took place. These were photovoltaic cells, thermo-pumps, wind and gas microturbines, power electronics including highly efficient motors and inverters, components of systems for automatic control and monitoring (SAC and M), etc.

---

[*] A_{eff} is the effective radiation surface area consisting of body bounding surface multiplied by the angular coefficients of radiation (shape coefficients). For parallel planes, we have $(A_{eff} = A_0)$.

New materials (including polyurethane foam, padding, reflective, selective and elec-tro-chromatic coatings, aero gel plates, and phase-changing materials) entered structural engineering and are successfully used in building envelopes, while other newly invented materials are still under testing. All this determines a new stage of the development of heat transfer and thermo-energetics, that is, the efficient use of primary energy resources. Except for the improvement in existing tools, the new trends in the field require the use of new, knowledge-based materials which can be adapted to the current energy hunger.

It seems useful to follow the evolution of the *energy-controlling function of the building envelope*[*] in order to illustrate those tendencies in the development of power technologies. A question arises of how the building envelope is related to the operation of the building energy system. To answer it, remember that there are two ways of supplying to buildings:

- Spontaneous supply from the surroundings (solar radiation and wind power uti-lization, entropy transfer—S)[†]
- Organized supply (via energy-logistic management of buildings—supply of natu-ral gas, oil, coal, or electricity)

Solar radiation possesses enormous penetration capability ($\lambda > 150$ nm) and a significant energy charge ($E(\omega) < 3.1$ eV). These properties make it a major factor in the design of the energy balance of buildings and building envelopes. As stated, it consists of

- High-energy photons (with $\lambda > 450$ nm), carrying about 7% of the total energy
- Visible photons with energy charge within the range ($1.7 < E(\omega) \leq 3.1$ eV) and energy share of 47%
- Thermal photons forming the infrared part of the spectrum ($\lambda > 760$ nm) with energy charge of $E(\omega) \leq 1.77$ eV and 46% share of the radiant solar energy

A significant part of the high-energy photons is absorbed by the ozone layer of the ecosphere (100 km above the Earth's surface) and by the air. Another portion of the solar energy scatters when absorbed by water vapors, aerosols, carbon dioxide, and other air components—nitrogen and oxygen.

Close to the ground layer of the atmosphere, between 17% and 40% of the direct radia-tion scatter reaching the ground objects.

It is noteworthy that the energy of solar photons reaching the Earth's surface and the enveloped buildings, respectively, depends on the geographical location and orientation of the object.

Table 5.1 shows data for the mean daily amount of solar energy per 1 m² horizontal sur-face area measured in the northern hemisphere. The fourth column specifies data for areas in Eastern Europe (41.8° northern latitude). It is seen that small and average buildings in

---

[*] Initially, the building envelope was used for protection against penetration, only. Later, during the Renaissance, aesthetic and harmony requirements were put forward, too. People became conscious of the energy-saving/blocking functions of the building envelope essentially in the second half of the 20th century, after the OPEC oil embargo.

[†] The spontaneous energy penetration into a building takes place via three mechanisms: diffusion (thermal conductivity), radiation, and convection (infiltration). Hence, architecture employed structural physics and found a technique of opposing energy fluxes by designing a building envelope operating as a thermal resis-tance. There are four envelope types: volume (Fourier) envelope—blocking diffusive transfer; capacitive envelope—blocking nonsteady transfer; reflective envelope—blocking radiative transfer; aerodynamic enve-lope—blocking convective leaks (occurring at dynamic pressure gradient or chimney effect).

TABLE 5.1

Mean Amount of Solar Energy Landing Daily on a Horizontal Surface Area, kWh/m²

Latitude L	60.4°	52.52	41.8	30.08	19.12
Direct	0.86	1.2	2.41	3.39	3.75
Diffusive	1.29	1.61	1.78	1.95	2.39
Total	2.15	2.81	4.19	5.38	6.14

this area, with an effective surface of up to 2500 m², can absorb and generate significant energy amounts during their annual exploitation. This energy flux, if incorrectly utilized, could be a significant energy burden hampering the regulation of the building's thermal and illumination comfort.

At the time of cheap energy carriers (the entire 20th century) the supply of buildings with energy from the surroundings was treated as an undesired external impact. Measures for its neutralization were undertaken designing the envelope as an "energy barrier or adiabatic insulation" to avoid or reduce energy penetration. Various envelope compact elements used in architecture are shown in Figure 5.13a and b. They suppress energy supply and transfer, increasing element thermal resistance to attain the necessary ergonomic minimum.

To function effectively as an "energy barrier," a compact element of the building envelope should be specially designed to reduce the expected energy fluxes (with specific energy spectra and orientation). This could be done by appropriate arrangement of thermal resistances along the gradient of the thermal potential—see for instance in Figure 5.13a where a unilateral energy barrier is shown. As seen, the reflective resistance 3 should be mounted first, followed by the volume resistance 1 and the capacitive resistance 7. However, when a bilateral "energy barrier" is designed (Figure 5.13b), the capacitive resistance 2 is mounted in the middle between two couples of reflective and volumetric resistances.

The idea of the passive regulation of energy fluxes directed to and from the building emerged in the 1980s during the subsequent stage of architectural evolution in the field. This was done by transforming the building envelope into an "energy filter" with a task

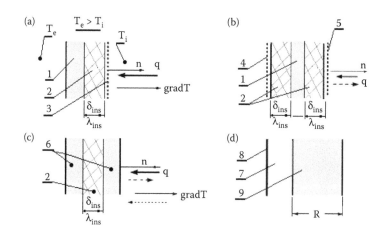

FIGURE 5.13
Arrangement of adiabatic and filter envelopes of a TDS. (a) unilateral heat barrier; (b) bilateral heat barrier; (c) bilateral filter; and (d) Trombe/Michelle wall: (1) capacity; (2) Fourier insulation; (3) reflective resistance; (4) wide-spectrum reflective resistance; (5) IR long-wave filter; (6) capacitive resistances; (7) air crevice (insulation); (8) window pane (UV/IR filter); and (9) heat capacity.

to shift in time or reduce (if necessary) the energy flux, while the appropriate time for this operation was chosen by the designer's decision. Note that if an envelope structural element is calculated, manufactured, and mounted on the building facade, it is expected to operate following a fixed program without a significant change of its functional characteristics during its whole life cycle (i.e., it is expected to be passive). The building envelope is designed to filter natural energy—to absorb or reflect it, if necessary. Most often building envelope elements are specially designed, that is, the energy receipt should drop in the time interval 10–16 h, and the energy should be stored for night consumption. For this purpose, building elements with "spectrally determined" characteristics are selected still at the design stage. Next, specific capacitive $R_C = 1/(C_v \cdot \omega_0)$ and volumetric $R_W = \delta_{Ins}/\lambda_{Ins3}$ resistances are arranged along the route of the diffusion flux (Figure 5.13c and d) or special filtering materials are arranged along the route of the radiation energy flux.

To change the heat capacity C, kJ/Km³ of envelope compact elements, one can use either classical materials, such as steel fiber reinforced concrete, ceramic plates, rock mass, or a combination of them with high "thermal mass," or materials which change their phase under a specific envelope temperature (about 290–300 K)—see Section 5.2.4.

The envelope filter functions properly when a suitable arrangement of the layers and their capacitive resistance are realized (as shown in Figure 5.13c for instance):

- Capacitive resistance ($\tau_{Charge} = 12$ h)—position 6
- Volumetric resistance—position 2
- Capacitive resistance (τ_{Charge} considering building habitation)—position 6

Quite often the envelope is built as a "wall of Trombe/Michelle" (Figure 5.13d), consisting of window pane (8), air layer (7), and steel fiber reinforced concrete plate (9). It makes use of the glass property to filter UVW^* and IRW and the thermal capacity of the steel fiber reinforced concrete. Filter layers are mounted in those envelope sections where the energy exchange is most intensive (the south-eastern, southern, south-western and western facades, as well as the building roof).

Hence, architecture has evolved in understanding the energy function of the envelope, passing through the following two stages (Figure 5.14):

- First stage—the envelope has been treated as passive with respect to external impacts (i.e., as a *barrier or filter*).
- Second stage—the behavior of the building envelope (its physical characteristics, respectively) can "accommodate" to the impacts of the surroundings, established by BASM and conforming to the requirement of maintaining the interior parameters within comfort limits, that is, the envelope behaves as an *intelligent membrane* (see Figure 5.15).

This stage was preceded in the 1990s by subsequent software simulations. These proved that there were significant resources in decreasing energy consumption for illumination and cooling (by 20%–30%) in residential and public buildings erected in areas with a moderate and hot climate. This could be done by using window panes that would let in sunlight through for interior illumination only. Thus, a prototype of a building with an intelligent envelope was designed in 2003 in the Lawrence Berkeley National Laboratory

---

* *UW*—ultraviolet and *IRW*—infrared wave.

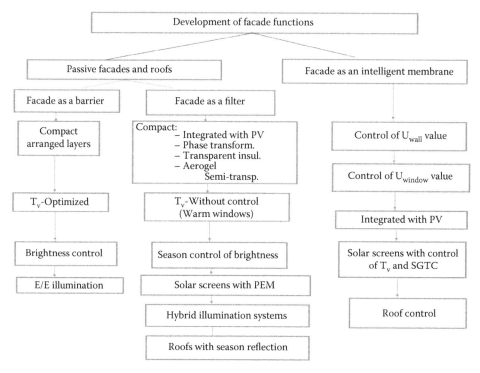

FIGURE 5.14
Scheme of the evolution of facade functions.

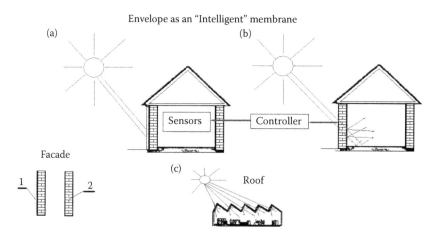

FIGURE 5.15
Scheme of a building envelope, type "intelligent membrane." (1—"spectral-sensitive layer" and 2—"transparent" carrying structure of the envelope.): (a) the facade membrane is "switched on"; (b) the facade membrane is "switched off"; and (c) the roof membrane is "switched off."

by a team headed by Dr. S. Selkowitz, an architect by profession. The facade comprised of window panes coated with thin ceramic electro-chromatic film. Their characteristics could be appropriately modified and controlled by a BSAM within a range conforming to the respective climate area (for the visible spectrum—transparency coefficient $0.05 \leq Tv \leq 0.6$ and coefficient of solar gains $0.09 \leq SHGC \leq 0.42$). The operation of the illumination system was controlled by BASM, measuring illuminance and the natural light flux landing on the windows. The internal light comfort was provided independently of the external illuminance via control of the degree of window shading and automatic switch on of fluorescent lamps.

We may assume that structural material science and technologies are close to the design of materials with "flexible" energy-transfer properties (thermal conductivity, for instance). They could have "spectral-sensitive" characteristics (not only transparency) and be pliable to total change under external programmed impacts (varying electromagnetic fields, in particular). Then, one could have a wider basis to model an envelope system as an "intelligent membrane" capable of accommodating to specific states depending on current needs. The development of building components with intelligent physical properties is a serious challenge to structural material science and requires the design of nontraditional nanotechnologies.

The following two approaches are essential in treating the problems thus outlined:

- Account for the integrity of the building envelope (the function of its components should be complex-protective, energy-saving, illuminating)
- Introduce the intelligent response of the component to external temperature impacts

We shall discuss in what follows some of the general techniques of design of thermal-insulating layers in energetic equipment-in building envelopes, in particular.

5.2.7 Calculation of the Thermal Insulation

During the execution of a specific energy project, specialists face the problem of maximal efficient usage of available energy resources. In fact, this is the problem of organizing the processes of conversion, transfer, and storage of primary energy with a task to get maximal benefit.

Assume that an energy-generating facility is treated as a TDS. Consider for this purpose combustion chambers, industrial ovens, fuel cells, heat engines, or some other devices.

Then, besides an adiabatic interface "TDS-surroundings," one should guarantee the recuperation of waste products such as combustion products and low potential steam (see Section 4.3). The efficient use of primary resources is an essential task of building projects, and the inevitable problem of building envelopes design arises. Consider an envelope consisting of compact and glazed facade components and roof plates. Then, besides purely energy-related problems, designers face also aesthetic and ecological problems.

The oldest traditional strategy of organizing the energy exchange with a TDS is the preparation of an adiabatic envelope as an absolute energy barrier. To fulfill its functions, a wall element of the adiabatic envelope should be specially designed accounting for the temperature gradient (see Figure 5.13a and b). When the temperature gradient does not change its direction (i.e., unilateral transfer runs), wall layers should follow the temperature gradient gradT, and their order should be capacitive layer, volumetric (Fourier) layer, and reflective layer (Figure 5.13a). When the gradient direction changes cyclically, the ordering of the different heat resistances along the normal is as that specified in Figure 5.13b:

- Reflective resistance against infrared waves (position 4)
- Volumetric (Fourier) resistance with δ'_{Ins} (position 1)

- Thermo-capacitive resistance ($\tau_{Charge} \cong 24$ h)—position 2
- Resistance reflecting the direct and diffusive radiation—position 5

Note here the specific problem of how to calculate the layer thickness and that of the volumetric (Fourier) resistance. The elementary logic of total insulation or thermal resistance tending to ∞ is incorrect, nonprofitable, and counter-indicative in some cases.

There are two types of restrictions on the excessive insulation of equipment and building components:

- Economical restriction or availability of "economically profitable or optimal heat resistance"–R_{ec}
- Physical restriction or availability of "critical thickness or critical heat resistance"

The economical restriction against excessive insulation is based on the social and personal investment interest. Assume an insulation with thickness δ_{Ins}. Then, minimum of the sum of initial investment (K_{Ins}) plus annual payments for energy supply (T_{En}) is sought. It has the form

$$NV_c = K_{Ins} + e^{-1} * T_{En}. \tag{5.33}$$

Here

- $K_{Ins} = f(\delta_{Ins})$ is a functional of the initial investment depending on the insulation thickness δ_{Ins}—curve 1 in Figure 5.16a
- $T_{En} = f(\delta_{Ins})$—function of energy supply expenses depending on the insulation thickness δ_{Ins}, $ (EU)/year (curve 2 in Figure 5.16a)
- e—discount factor of the current expenses at the moment of making the initial investment[*]
- NV_c—net present value, $ (EU) (curve 3—Figure 5.16a)

The insulation thickness ($\delta_{Ins} = \delta_{ec}$) minimizing NV_c is found by means of an optimization procedure (Figure 5.16a). It is called economically profitable thickness of the insulation.

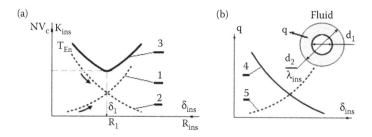

FIGURE 5.16
Restrictions imposed on the envelope dimensions. (a) Optimization of the mean expenses NV_c and (b) specification of envelope critical thickness: 1—initial investment $K_{Ins} = f(\delta_{Ins})$; 2—annual energy expenses $T_{En} = f(\delta_{Ins})$; 3—NV_C—investment present value; 4—$q = f(\delta_{Ins})$; and 5—$Q = f(\delta_{Ins})$.

---

[*] The discount factor reflects future payments.

If $\delta_{Ins} \geq \delta_{ec}$, the object is excessively insulated. Laying insulation with thickness exceeding δ_{ec} is economically unjustified, and it yields unreasonable high initial investment costs (thus making the idea of attaining absolute adiabatic insulation unrealistic).

To save investments, one can find the thickness δ_{ec} using the indices and classes of energy efficiency. This approach is highly simplified and guarantees good communication between designers and investors. According to this method, δ_{ec} is found as

$$\delta_{ec} = \frac{\lambda_{Ins}}{U_{ref}} * IEE^{-0.5} = \lambda_{Ins} * R_{ref} * IEE^{-0.5}, \tag{5.34}$$

where

- U_{ref}, W/m²K—reference value of the thermal conductance of the energy standard[*] legally regulated, depends on type of climate (for temperate—continental climate, currently may be used $U_{ref} = 0.4$ W/m²K)
- R_{ref}, m²K/W—reference value of the thermal resistance of the energy standard legally regulated (actually $R_{ref} = 2.5$ m²K/W is recommended)
- λ_{Ins}, W/mK—thermal conductivity coefficient of the used insulation material
- IEE_{Env}—index of the energy efficiency specified by the investor depending on the adopted Class of energy efficiency: for instance, for class a—$IEE_{Env} = 0.625$, for class B—$IEE_{Env} = 0.875$, for class C—$IEE_{Env} = 1.125$, etc.

Insulation design requires that the investor should declare the class of energy efficiency, only for instance class A, B, C, etc.[†] Then, the index of the envelope IEE_{Env} is found (using a normative table of correspondence). Figure 5.17 presents the values of δ_{ec} depending on IEE_{Env}(or CEE) for reference values of the conductance coefficient within the range $0.2 \leq U_{ref} \leq 0.6$ W/Km² if $\lambda_{Ins} = 0.03$ W/mK.

These data show that for the high class of energy efficiency—A, δ_{ec} attains high values (0.07–0.12 m). For classes lower than B, the thickness of an economically profitable heat insulation is within ranges $0.03 \leq \delta_{ec} \leq 0.08$ m. As found, it is not economically expedient for δ_{ec} to exceed $\delta_{ec} = 0.08$ m in mass building of structures.

Buildings with class of energy efficiency D ÷ G, whose insulation is thinner than 0.05 m, should be recovered (subjected to architectural-structural "healing").

The second type of restrictions concern axisymmetric objects (pipes), in which a heat carrier with temperature T_1 flows—Figure 5.16b. When a pipe whose insulation thickness is δиз releases heat to the surroundings (with temperature T_2), two opposite effects occur:

- The specific heat flux $q(\delta_{Ins})^{\Downarrow}$ decreases[‡] with the increase of δ_{Ins}, due to the increasing resistance R_{Wall}
- On the other hand, the heat flux to the surroundings $Q(\delta_{ec})^{\Uparrow}$ increases with the increase of δ_{Ins} due to the increase of the effective cooling surface of the insulated pipe[§]

---

[*] A virtual building without glazed elements, with volume equal to that of the designed building and with spherical shape is taken as a standard (i.e., its surface area is minimal). This is the so-called *spherical standard*.
[†] You can find the details in the book *Energy Modeling and Computations in the Building Envelope*, CRC Press (2015).
[‡] $q = (T_1 - T_2)/[\ln(1 + (\delta_{Ins}/d_{Pipe}))]/2\lambda_{Ins}$.
[§] $A_{eff} = 2\Pi L_{eff}(d_{Pipe} + 2\delta_{Ins})$.

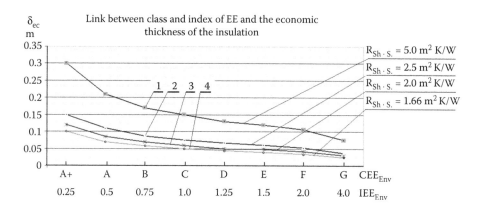

FIGURE 5.17

Maximal admissible thickness of the thermal insulation. (Values of δ_{ec}, calculated for different $R_{ref} = R_{Sh \cdot Sec}$ depending on the class CEE_{Env} and index IEE_{Env} of the energy efficiency of an envelope manufactured from an insulating material with $\lambda_{Ins} = 0.03$ W/mK.)

It is obvious that a limit (critical) value of δ_{u3} exists, and its exceeding is senseless and physically ill-founded.

When we have $0.5U_{In}(d_{Pipe} + 2\delta_{Ins})/\lambda_{Ins} \leq 1$, heat exchange decreases. In the opposite case, heat flux increases, and this seems paradoxical outside the scope of the above interpretation. The limit thickness δ_{Cr} can be accessed via the following inequality:

$$\delta_{Cr} \leq \frac{\lambda_{Ins}}{U_{Out}} - 0.5d_{Pipe}.$$

When $\delta \gg \delta_{Cr}$ heat exchange increases and the use of insulation is senseless. In the opposite case, the insulation functions accordingly.

5.3 Control Questions

1. Which is the basic energy carrier in solids and liquids?
2. What are Einstein's waves? What are Debye's waves?
3. How many types of phonons do exist? Which is the class of quasi-particles that the phonons belong to?
4. What are isotherms? How many isotherms may pass through a point in space?
5. Define the vector "gradT" with respect to magnitude and direction.
6. Is the gradient large or small when two isotherms are very close to each other?
7. How many types of thermal resistance exist?
8. A vector characteristics of transfer called "transfer density $\bar{q}$" exists. What does it mean?

9. What does the total energy transfer law say?

10. What does Fourier law of heat transfer state?

11. What is the analogy between the laws of Fourier and Ohm?

12. What is the physical meaning of the coefficient of thermal conductivity λ? How does temperature affect λ?

13. What is the relationship between λ and Ω (coefficient of electro-conductivity)?

14. Which are the basic physical models of heat conduction in solids?

15. Which are the similarities and differences between Kaganov's model and that of Geier/Kumanasi?

16. Which area of a TDS is the A. Majumder model applicable to?

17. How many types of heat exchange exist at the macroscopic level?

18. How is the thermal resistance of thermal conductance defined? How does a large value of R affect the heat flux?

19. How can one calculate the coefficient of thermal transitivity U_W for multilayer walls? What is a reference value of U_{ref}?

20. How does the thermal capacity C_v vary with temperature T? How can one explain this variation considering low temperatures?

21. What are the conditions of bilateral operation of a thermo-capacitive element?

22. What does the efficiency of a thermo-capacitive resistance depend on?

23. How can one find the thermal resistance of a TDS whose working fluid is water, but whose control boundaries are very close to each other, and the viscous forces hamper water flow?

24. How does heat transfer take place when the area occupied by the fluid expands?

25. What does the law of Newton–Richman state?

26. What does the thickness of the boundary layer δ_{fl}^* depend on and how does it affect the coefficient of heat release U_{Out}?

27. Which are the standard values of U_{Out} and U_{in} for a vertical wall?

28. How does the thermal resistance of an air pocket depend on the layer thickness? How does it depend on the fluid density?

29. What are the reasons of painting compact walls in different colors?

30. What does the principle of superposition of heat fluxes state?

31. What is the TDS-stored energy converted in?

32. What is the function of the heat accumulator?

33. How do enthalpy accumulators operate? What are the properties of the working medium?

34. Which are the types of latent accumulators used in energy equipment? How do they operate? Which are their thermodynamic cycles?

35. Outline the areas of application of heat accumulators.

36. What are heat-accumulating elements in building envelopes used for?

37. Besides high C_v, which are the general properties that a heat-capacitive working medium should possess?

38. What is the behavior of a heat-capacitive element located in a steady temperature field?

39. How does that element react to temperature harmonic variation?

40. Which are the frequency temperature oscillations that pass through a heat-capacitive resistance?

41. Which temperature oscillations are absorbed by a heat-capacitive resistance? Why does this happen?

42. How can a heat-capacitive resistance function as an energy filter?

43. Can heat be transferred through vacuum?

44. What is the similarity between photons and phonons?

45. What is the energy of a single photon? Which quantity specifies the energy of an orbital occupied by a photon?

46. How can the number of photons of an energy orbital be regulated?

47. What does Bose's distribution function of include?

48. What does the Plank integral indicate?

49. What does the Stefan–Boltzmann law state?

50. Which bodies are perfectly black? Which bodies are perfectly transparent and perfectly white?

51. What does the Lambert law state?

52. How is the coefficient of radial heat release U_R defined?

53. How can one regulate radiation heat exchange?

54. Which photons are components of solar radiation?

55. What is the mean amount of solar energy per 1 m² in our latitude?

56. What is the share of direct radiation in the total solar radiation?

57. Which heat insulation is known as adiabatic one? How are the insulation layers ordered so as to attain insulation adiabatic operation during bilateral supply of the thermal gradient?

58. Which insulation is called passive filter? What is the order of its layers?

59. How is the insulation thickness determined depending on the class of energy efficiency?

60. Why is the use of heat insulation with thickness $\delta_{Ins} > \delta_{ec}$ nonprofitable?

61. What is the result of tube insulation if the insulation external diameter exceeds the critical one?

62. What is the function of entropy accumulators and how can one increase their capacity?

63. How do the latent heat accumulators operate? Which are the requirements to their working medium?

64. Which are the types of recuperative heat exchangers used nowadays?

65. How do mixing heat exchangers operate? What is the area of their application?

6

Energy and Indoor Environment

6.1 The Necessity of Designing a Comfortable Indoor Environment

Indoor environment is a complex term, comprising the physical space of occupied areas in buildings or vehicles and habitation conditions. Assume that the best working and indoor conditions are provided in a comfortable indoor environment, having a favorable effect on inhabitants–dispatchers, maintenance, analytical staff (managers, designers, accountants, auditors, etc.), and regular inhabitants. The available optimal working conditions stimulate staff activity and productivity and decrease the number of operational errors, morbidity rate, and absence because of ill health.

It is found that people spend 45–90 min daily to travel to and from office, regardless of whether they live in Sofia, Torino, or Mumbai. Practically, an active person spends about 18%–20% of the time in vehicles or train, bus, and subway stations or airports. When the indoor environment is favorably arranged and passenger friendly, it reduces the stress of travelling. On the other hand, if the transport indoor environment is "hostile," it increases stress and the risk of illness, operators' faults and catastrophes resulting in damages and loss of life. The organization of an inhabitant friendly indoor environment is an obligation of the *operational* staff of buildings, transport, and infrastructure. This is the main task, priority and responsibility of building, industrial, and transport *managers*.

A favorable and comfortable indoor environment can be attained employing two architectural-engineering technologies that complement each other:

- Passive technologies
- Active technologies

The first type exploits the functional properties of the envelope of an object (building or vehicle) to filter environmental impacts such as solar radiation, wind, humidity, and aerosol emissions. Once designed and laid, the envelope operates as an insulation, saving energy needed to provide a friendly indoor environment.

The second type uses energy generated in central or local power stations.* That energy is needed by the devices to change the parameters of the indoor environment, offering thermal, illumination, and air comfort.

---

* Central electric power stations (thermal, water, and nuclear) and central heating stations (thermal, geothermal, and nuclear).

6.2 Energy Needed for Thermal Comfort in an Inhabited Area

Energy, needed for people's activity and working capacity is generated by dissimilar processes, that occur in the human body accompanying the process of metabolism. Their normal operation requires the maintenance of a constant human body temperature T_{HBody} within the range $36.7 \leq T_{HBody} \leq 37.2°C$. As known from the first principle of thermodynamics, the "chemical energy—work" conversion in the human body is accompanied by release of "waste" heat. Part of that heat is used to maintain the body temperature. The other part is lost as energy waste through the human skin and lungs (see Table 6.1).

Body thermal comfort is attained if there is a balance between the incoming and outgoing heat fluxes:

$$Q_M \mp Q_C \mp Q_R \mp Q_{Con} - Q_E = Q_M + Q_S + Q_L = \pm \Delta Q = 0. \tag{6.1}$$

Here

- Q_M—heat flux due to metabolism
- Q_C, Q_R, Q_{Con}, and Q_E—heat fluxes due to convection, radiation, conductance, and evaporation
- Q_S and Q_L—sensitive (due to the temperature gradient) and latent (due to moisture evaporation) heat fluxes

Human body temperature T_{HBody} is controlled by a system for biological regulation, consisting of the following mechanisms:

- Increase of skin temperature and activation of sweating for $\Delta Q > 0$
- "Goose-flesh" and activation of the muscle metabolism (shudder) for $\Delta Q < 0$

If this system does not operate effectively, the body superheats or overcools, and the result can be lethal in both cases.

One should take into account that metabolism and the released heat depend on a number of factors, including human activity, weight, gender, age, clothing, and emotional state. To illustrate that, Table 6.2 specifies the mean values of the metabolic heat flux for four types of activity and under air temperature T_a within the range $15 \leq T_a \leq 35°C$.

Regardless of the large number of factors which affect thermal conditions, engineering practice uses only four factors to guarantee zero thermal balance (i.e., condition of thermal comfort). They guarantee and regulate surroundings comfort, i.e.,

- Indoor air temperature T_a
- Airflow velocity w_a

TABLE 6.1

Type of the Heat Fluxes, Emitted by a Human Body

Through Skin			Through Lungs	
Radiation	Convection	Evaporation	Heating	Evaporation
42%	26%	18%	7%	7%

TABLE 6.2

Relationship of the Total Heat Flow, Transferred from the Human Body from Temperature of the Air and the Activity of the People

	15°C	20°C	25°C	30°C	35°C
At rest	145 W	116 W	93 W	93 W	93 W
Easy work	157	151	145	145	145
Medium work	208	203	197	197	197
Hard work	290	290	290	290	290

- Indoor radiant temperature T_R
- Indoor relative humidity φ_a

"Friendly" surroundings are attained by a favorable combination between these four factors.[*]

Figure 6.1 shows the bioclimatic graph of V. Olgyay as an illustration. It is received from people doing sedentary work (reading, writing, working on a computer, dressed in a suit with isolation level 1 Klo and acclimatized for the climatic conditions of Middle Europe).

Besides the comfort operative values in the occupied area, the bioclimatic graph also reveals the mechanisms of attaining comfort outside the hatched areas in the figure. Thus, the so-called "standard approach in attaining thermal comfort" is outlined for $w_a = 0$ and $T_a = T_R$, namely

- In areas under the "*hatched*" area, thermal comfort can be also attained at a temperature lower than 18°C, if the requirement $T_a = T_R$ in the indoor area is violated via radiation with photons belonging to the infrared spectrum (it is desirable that the radiation specific flux exceed 70–75 W/m^2).

- In areas over the "*hatched*" area, thermal comfort can be attained by violation of the condition $w_a = 0$, that is, by activation value of the air velocity w_a is functionally related to the value of T_a, of the forced convection within the inhabited area (the "comfort" but it should not exceed 0.2–0.4 m/s.) Yet, for higher convective velocities [over 2 m/s], most people experience discomfort, i.e., have a feeling of an "air draught").[†]

The air relative humidity φ_a in the comfort area of the plot changes within limits $15\% < \varphi_a < 75\%$. When φ_a is below the comfort limit, people experience dryness (dry skin syndrome). There are no data in the technical literature proving that stays in areas with low humidity is injurious to health. If φ_a exceeds the upper limit of discomfort, the latent heat exchange between the human body and the indoor environment is hampered. This occurs when high relative humidity ($\varphi_a > 90\%$) is combined with high air temperature T_a, $T_a > 30$–32°C, for instance. A long stay in such an environment yields heavy perspiration, dehydration, and an increase of body temperature. As a result, a person's working capacity sharply drops and inversion in the blood flow in the circulatory system may occur.

---

[*] The plot of the field of the possible combinations between those factors is called bioclimatic graph proposed in 1963 by the Italian architect V. Olgyay.

[†] The alternative approach is when the indoor radiant temperature T_R is maintained with air temperature T_a by means of cooled walls or floor surfaces. This has a risk of condensation of moisture on the building structures and the appearance of mold or fungus.

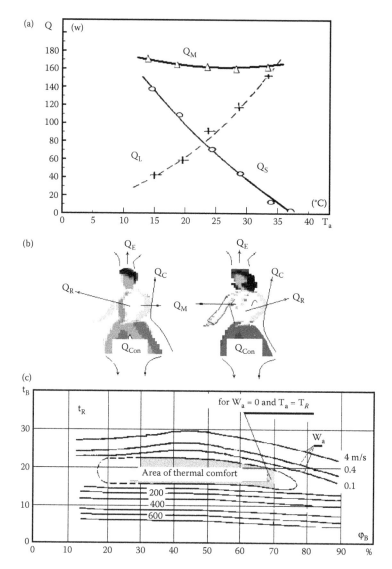

FIGURE 6.1
Scatter of metabolic heat: (a) dependence of the metabolic heat fluxes on air temperature T_a, (b) structure of the metabolic heat exchange, and (c) bioclimatic chart.

It has been learned that relative to thermal comfort, despite the change of indoor air temperature T_a, the mechanisms of "correcting" the environmental parameters in order to maintain thermal comfort within admissible limits also include a change of air velocity W_a and indoor radiant temperature T_R.

Hence, a classification of the control systems for thermal comfort is proposed on such a basis (see in Figure 6.2). It includes control of air conditioning systems[*] or heating systems only (without active control of air humidity φ_a).

---

[*] In modern sense, air conditioning systems were designed after 1902 when the American engineer Willis Carrier developed psychometric diagrams and applied an original technology to actively control air humidity φ_a.

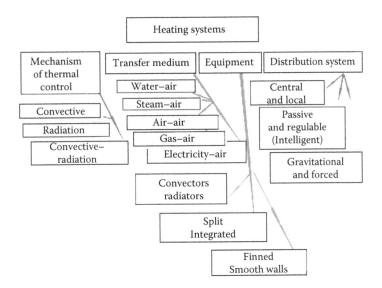

FIGURE 6.2
Classification of the systems for maintenance of thermal comfort (heating and air conditioning).

Yet, the rich variety of the control systems for thermal comfort can be hardly covered in the present section due to the vast scope of the subject.

As seen in Figure 6.2, the systems of thermal comfort regulation are divided into three basic groups with respect to the mechanisms providing comfort in the inhabited area

- Convective systems. The necessary local conditions for thermal comfort in the inhabited area are created by convective heat transfer and change of the air temperature T_a.
- Radiative systems. Comfort via those systems are attained mainly by change of the radiant temperature T_R of the heaters located around the inhabited area. Air temperature T_a is generally kept constant.
- Convective–radiative systems. Temperature T_a as well as temperature T_R vary in the conditioned area.

Considering the working fluid in the distribution network, the systems are subdivided into

- Air-to-air
- Water-to-air
- Steam-to-air
- Electricity-to-air
- Gas-to-air

With respect to heat transfer, they are classified as

- Heating systems using natural convection
- Heating systems using forced convection

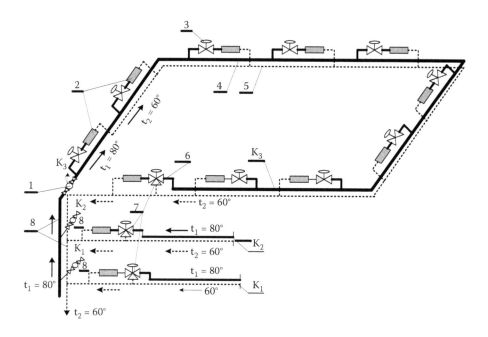

FIGURE 6.3

Perimeter system with individual heat-meters in each loop: 1—heat meter; 2—heaters (convectors or radiators); 3—two-way thermostatic valve; 4—horizontal supply pipe ($T_1 = 90/80°C$); 5—horizontal return pipe ($T_2 = 70/60°C$); 6—by-pass; 7—three-way valve; 8—risers-up/down.

Systems providing thermal comfort (air conditioning or heating) reflect the specificity of the objects involved. Hence, engineering offers a number of systems which differ from each other with respect to the internal arrangement of heat distribution. This will be illustrated by several examples owing to *structural engineering and transport systems.*

In residential, public, and commercial buildings where the basic energy needs are determined by current environmental impacts, so-called "perimeter systems" are designed. Their specific feature is that heat needed to provide an acceptable thermal indoor environment is supplied at spots close to the building envelope (heaters are often mounted under the windows) and at other proper locations in the building *perimeter area.*[*]

Figure 6.3 shows the scheme of a *perimeter heating system* type "water to air." It is organized on each floor and has a hydraulic loops consisting of two horizontal distribution pipes (supply 4 and return 5). The first pipe feeds heating devices 2 with hot water (90–80°C), while the second one collects the water used and cooled down to 70–60°C.

The number of horizontal loops corresponds to the number of floors.[†] The supply of hot water to and collection of cooled water from all heating devices is performed by leveling the hydraulic losses, employing a properly arranged pipe system, type "Tichelmann." A heat-meter 1 is mounted at the intake of each horizontal system. It reads the heat energy used within the horizontal loop. At the same time, the scheme admits individual regula-

---

[*] The so-called "central space air conditioning/heating systems" are used in buildings where energy demand is distributed evenly throughout the building volume. They are "water to air" type employing air forced convection.

[†] If there are several flats on a floor, the subsequent number of hydraulically independent analogs should be arranged.

tion of the heat released by each heating device by means of two-position thermostats 3. A three-way regulation valve 7 is fixed to the last heater mounted along the hot water flow. Thanks to the bypass 6 between the supply pipe 4 and the return pipe 5, hydraulic stability of the pipe network is guaranteed when all thermostats along the loop are closed.

All horizontal loops ($K_1 - K_6$) are fed by a riser-up 8 connected to the thermal generator/boiler (or heat exchanger) located at the building energy center (the boiler room or the room where the heat exchangers controlling the energy consumption are mounted). To realize a full heating cycle within the system, the used heat carrier is applied with decreased temperature T_2 ($T_2 = 70-60°C$) and is collected in the riser-down 8 and introduced into the boiler/or cogenerator for repeated heating to $T_1 = 90-80°C$.

The heaters and pipe elements are properly calculated to reach the *maximal power demand* for the conditioned area where they are installed. Then, the automatic control system (SAC or BSAC and M) comprising individual thermostats 3, three-way valves 7, and the automation of the circulation pumps and the boilers, supplies the area with currently needed power, only.* Thus, economy of the primary fuel is realized and operational efficiency is achieved.

Systems providing thermal comfort in *vehicles* should more fully correspond to the specific features of the occupied area, as compared to similar stationary systems employed in structural engineering. Those features are

- Regime of human occupancy of vehicles
- Vehicle thermal-indoor properties, including dimensions of passenger compartment and insulation thermal characteristics
- Character of the indoor and external heat loads

The acceptable level of indoor air conditions (*ALIAC*) of vehicles is specified as dependent on the time of travel of passengers. Air temperature T_a in the range $22.5 \leq T_a \leq 24.4°C$ should be maintained in vehicles for passenger travels *longer than 1 hour* (according to the European standard, $T_a \geq 21°C$ and $6° < T_{cyx} - T_e \leq 5°C$ in summer). The relative humidity should be maintained within limits $55 \leq \varphi_a \leq 60\%$. Vehicles where passengers *spend a short time* (10–20 min) allow for large variations from the specified values.

The model of energy consumption of vehicles is designed depending on the specific meteorological conditions of the geographic area (i.e., whether the climate is moderate, continental, tropic, sub tropic, sea, or polar). The model contains a detailed description of the basic components of energy expenses with respect to type and time. When these data are lacking, one should enclose only a list of their share distribution with respect to type.

To illustrate the above considerations, Figure 6.4 shows the distributions of the energy flows in a bus with 50 seats rolling under environmental temperatures of $T_a^{Dry} = 27°C$, $T_a^{Wet} = 19°C$ and with a speed of 60 km/h. Its insulation consists of solid elements with thickness $0.025 \leq \delta_{Wall} \leq 0.038$ m, "sandwich" type, and double glazed windows. The total conditioning power needed is 24 kW, while the relative specific "contributions" of the thermal loads are illustrated in Figure 6.4. They are heat exchanged through the windows (44.6%), heat released by the passengers (25.9%), heat losses from infiltrated outdoor air (16.1%), total energy needed to make-up the infiltrated external air (24.3%), and total heat gains through the walls (14.4%). That model of energy consumption is designed employing computer simulation. Yet, it can also been developed performing physical experiments in the laboratory or "in situ."

---

* Although building systems are growing more and more "intelligent," maintenance and control performed by the "power supply" staff are of crucial importance for their efficiency.

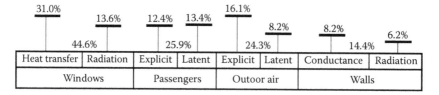

Heat transfer	Radiation	Explicit	Latent	Explicit	Latent	Conductance	Radiation
Windows		Passengers		Outoor air		Walls	

FIGURE 6.4
"The model" of energy consumption of a vehicle.

"The model" of energy consumption of a vehicle is essentially important for the synthesis of the system controlling the vehicle thermal comfort.

Considering vehicles with a structure of energy model similar to that shown in Figure 6.4, *two types of systems* providing thermal comfort are considered with respect to their area of application

- Systems for intercity transport with radius of operation exceeding 40 km (see in Figure 6.5)
- Systems for city transport (see in Figure 6.6)

A system maintaining thermal comfort in busses for intercity transport is built by two subsystems

1. Subsystem supplying cooled air (Figure 6.5b)
2. Subsystem supplying warm air (Figure 6.5c)

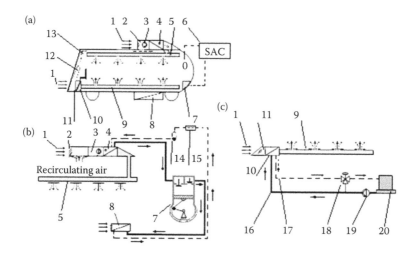

FIGURE 6.5
A system for control of the thermal comfort in an intercity bus: (a) general arrangement of the equipment; (b) chilling system; (c) heating system: (1) intake of fresh outdoor air; (2) diffuser; (3) recovery section with supply fan; (4) chilling section with cooling coil/evaporator; (5) distributing air duct for cold air; (6) SAC; (7) compressor aggregated to ICE; (8) condenser; (9) duct for hot air; (10) heat exchanger; (11) hot-air distributor box; (12) hot air; (13) exhaust air; (14) receiver for refrigerating aggregate; (15) expansion valve; (16) supply pipe for hot antifreeze; (17) return pipe for antifreeze; (18) three-way valve/thermostat; (19) circulation pump; (20) ICE-cooling system.

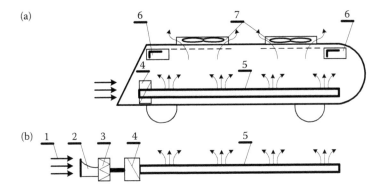

FIGURE 6.6
Heating installation of a city bus: (a) vertical cross section of a city bus; (b) Linear scheme of the duct system: (1) intake of fresh external air; (2) diffuser; (3) filter; (4) heat exchanger; (5) hot air distribution duct; (6) exhaust with flap; (7) exhaust fans.

Each of them is controlled automatically or from the driver's control panel (the time of system automatic switch-off is called dead zone). When the air temperature T_a in the compartment exceeds a specific value (for instance, $T_a > 27°C$) a cooling system is actuated automatically or by the driver. It comprises (see in Figure 6.5b)

- Intake diffuser 2
- Recovery section equipped with a supply fan 3
- Chilling section and cooling coil 4
- Distributing air duct for chilled air 5
- Refrigerator (compressor 7 aggregated to the internal combustion engine [ICE] of the vehicle, condenser 8, expansion valve 15, receiver 14, and cooling coil/evaporator 4)

Outdoor air is drawn through the intake diffuser 2 located on the compartment roof. It is mixed with recirculation air drawn from the compartment by fan 3 in the recovery section. The proportion between fresh and recirculating air is regulated automatically or manually by means of a control flap. The air mix thus obtained passes through chilling section 4 and enters the distribution duct 5. The velocity of air exhaust is set manually by the passengers to provide thermal comfort. The chilling system is switched off when the air temperature T_a remains lower than the desired value ($T_a < 24°C$, for instance) for a long time. Systems for compartment cooling or heating are switched off during transitory seasons, when the external meteorological conditions allow this. In these conditions they can be used to supply fresh air necessary for the operation of the ventilation system.

When air temperature T_a decreases below a specified limit (for instance, $T_a < 18°C$), a heating subsystem switches on.

It consists of the following basic elements (Figure 6.5c):

- Heat exchanger (heating coils) 10
- Duct for hot air 9
- Cooling system of the ICE consisting of a circulation pump (19), three-way or thermosatic valve (18) and heat exchanger 10

Outside air intake 1 is through the distribution box 11, located on the front panel of the vehicle. It is heated by heat exchanger 10 and distributed into two flows:

- Flow toward the windshield pane and around the driver
- Flow toward the compartment where the duct for hot air 9 is mounted

Energy waste of ICE is used during transition seasons and in winter. Thus, efficiency in the use of the primary carrier is increased. This is done using the *cooling system of the ICE* consisting of supply 16 and return 17 pipes, three-way thermostat 18, and the circulation pump (19). Depending on the power needs, heat owing to engine cooling is released either to the surrounding (external) air (the cold source) or distributed in the compartment to maintain thermal comfort. Heat exchanger 10 of the heater, which regenerates waste heat, has a two-section or multi-section configuration. Then, one section operates in a mode of continuous cooling of the ICE while the other one is activated only if the compartment or driver's cabin needs heating.

When the vehicle runs within a populated area, only, passengers would occupy it for a short time and its compartment would be densely "occupied." Then, air parameters should be lower: $T_a \rightarrow 16°C$ during heating and $T_a \rightarrow 19–22°C$ during cooling. The heating system should guarantee increased rate of fresh air delivery (for instance, 16–30 times larger volume of fresh air supplied per hour).[*]

A typical scheme of a vehicle driven in built-up areas, only (city drive) is shown in Figure 6.6. The scheme is *identical* to that shown in Figure 6.5c, but it is intensified using roof fans 7,[†] which take in polluted air from the under-roof space of the compartment. Carbon dioxide exhaled by passengers (which is heavier than the air mix) is removed by opening the doors when the vehicle is at rest.

The above schemes are applied in buses but similar schemes are also used in railway transport units (subway carriages, trams, and railway carriages). The latter are designed considering the time of passengers' stay in the vehicle.

Railway carriages are classified as follows:

- For short time transport (metro-carriages, lift cabins, trams for in-city transport with radius of up to 30 km)
- For medium time transport (from 45 min to 3 hours—intercity transport)
- For long time transport over 3 hours (inter-district transport)

The indoor environment conditions of railway transport are identical to those of bus transport ($T_a = 21°C$ and $\varphi_a = 55\%–60\%$). Departure from those norms is permitted only in cases when passengers are able to stand short travel discomfort (5–15 min). The thermal input is calculated accounting for peak travel hours. The following thermal comfort is provided in compartments of railway cars, considering the short-time stay of passengers:

- Heating in winter and in transitory periods
- Annual ventilation (30–40 times increased air circulation per hour)[‡]

---

[*] The system of thermal comfort maintenance should guarantee a stable rate of fresh air delivery, considering any driving regime (speed, stop, and in traffic) and any season (including driving in winter with closed doors and windows).

[†] The rates of heat-air delivery and air conditioning should be guaranteed for any driving regime (speed, stop, driving in traffic) and season (including driving in winter with closed doors and windows).

[‡] The concentration of many people in a limited space is a serious precondition for the spread of air-borne diseases.

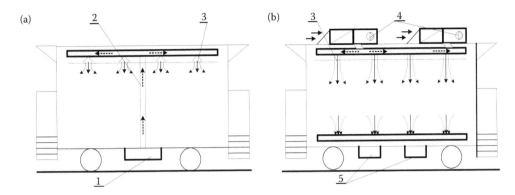

FIGURE 6.7
Systems for control of the thermal comfort of railway carriages: (a) electricity-to-air heating system; (b) central air suppliers mounted under the carriage: 1—air handler; 2—duct distribution system; 3—exhaust nozzles for hot air; 4—evaporation/ventilation group; 5—condenser/compressor group.

The requirements of air quality are higher in those passenger compartments—where air ozonation is performed and CO_2, NH_3, moisture, perspiration odor, etc. are removed by ventilation.

In railway vehicles covering short and medium distances, electric and electro-air heating systems prevail (see in Figure 6.7a), and central air suppliers mounted under the carriage are used (see in Figure 6.7b).

Hermetically sealed carriages have been increasingly used in passenger trains in recent 15–20 years. They are entirely air conditioned and provide comfortable travel. The indoor environment conditions are close to those in planes and luxury ships. The air conditioning systems have two air handler units with a common control panel (each of them can autonomously provide up to 75% of the energy needed for carriage conditioning). The carriage conditioning packages are energy supplied by a central generator (alternator) or inverter. Split systems are used where

- Evaporation/ventilation group (position 4 in Figure 6.7b) is mounted on the roof or in the attic
- Compressor/condenser group 5 is mounted under the carriage

A new feature of the air conditioning of railway cars is the use of a modern system for automatic control and monitoring. Analog thermostats are replaced by microprocessor consoles. The system for automatic control is developed on the basis of digital control (DDC). Besides indoor environment parameters control, it carries out diagnostic and service tasks.

Another important feature of modern carriage air conditioning is that the so-called nonecological refrigerants R12 and R22 are replaced by ozone friendly compounds such as HFC194, HFC 407C, HFC 410A, or HFC 410C. The used compressors are hermetically sealed, type "scroll," centrifugal or "in tandem."

When selecting appropriate equipment for air conditioning systems, one should control whether they are compact and have small weight, can be easily assembled/dismantled during travel, are hermetically sealed and reliable, and can be easily serviced and cleaned. Besides, they should be corrosion resistant and should not generate noise and vibrations, especially in the passenger compartments. The air conditioning systems of hermetically sealed cars should operate during travel and at rest.

6.3 Energy and Clean Air

6.3.1 The Necessity of Clean Air in the Occupied Area and Methods of Its Supply

As is known, oxygen is vitally important for the operation of energy equipment and most of all, for human life. It is a necessary oxidizer in dissimilation processes running in the human body at the cell level, when food is converted, heat is released, and physiological work is done. At the same time, continuous metabolism requires the removal of the oxidation waste-carbon dioxide (CO_2) and water from the cells and from the entire body. The energy produced should be properly distributed or "released."[*] Air quality is of crucial importance for the progress of the physiological processes in the human body. Hence, air quality in habitation areas (residential, public and business buildings, factories, etc.) predefines human activity and health.

As far as the air quality in buildings is formed spontaneously or under control depending on the environmental effects (solar radiation, winds, infiltration, and contaminants), it is affected in passenger compartments by the operation of the thermal engine and its systems, as well as by the concentration of passengers and loads (animals, chemical compounds, building materials). The surrounding air is a gas mix of oxygen (21%), nitrogen (76%), carbon and nitrogen dioxide, water vapors, aerosols, etc. Some of its components, including CO, Cl_2, SO_2, HS, CS_2, NO_x, and radon, are aggressive toward human tissues. They are hazardous poisons, which interact with the mucosa of eyes and respiratory organs and with nonprotected skin areas. At small concentration and short contact, they cause inflammation, and at large concentration and long contact—flesh wounds, allergy, and other more serious injuries.

Air ingredients such as CO_2, NO_2, water vapor, and methane (CH_4) are neutral with respect to human tissues. They are not poisonous, but they decrease the oxygen concentration in the air, and their presence in the air is undesirable.

Carbon dioxide CO_2 is a natural product of dissociation running in biological organisms. It has no color and odor, but is heavier than air and concentrates over the ground and close to the Earth's surface. Carbon dioxide is also released by the combustion chambers of boilers and engines during their operation. It may penetrate the operator's cabin and the passenger compartment through gaps in the insulation or diffusers of systems for thermal comfort control. This may happen at a traffic light stop, in a traffic jam or in closed parking lots, resulting in air contamination. The concentration of carbon dioxide depends on the number of occupants of the indoor area or vehicle passengers. It varies depending on the flux of the incoming fresh air[†]. Its acceptable concentration should not exceed $1.8 \text{ g}/\text{m}^3$ (or 1000 ppm).

Quite often, during short engine run time and due to incomplete fuel oxidation, a significant amount of carbon monoxide CO is released, which is lighter than air, without color and odor, but quite toxic. Concentration of CO higher than 15 ppm is considered to be hazardous to health. Carbon monoxide is a very active gas and is easily absorbed by human organisms. People having inhaled CO suffer from headaches, nausea, and weakness.[‡] Sites

---

[*] From an ecological point of view, man is a biological species, at the top of the food chain. Consuming oxygen, the human releases CO_2, H_2O, and excrement.

[†] Studies of 100 airplanes were performed and a concentration of 1500 ppm was found in airplane passenger compartments. 5000 ppm (or 0.5%) was adopted as a limit norm by the FAA.

[‡] They should be immediately hospitalized in a specialized clinic.

with potential hazard of CO release are closed parking lots, docks, crossings with traffic lights, tunnels, and traffic jams. There, air contamination is avoided by designing an effective air ventilation system.

When *microbiological organisms* (*bacteria* and *viruses*) are present in the air, passengers often get flu, pneumonia, and other lung diseases.[*] The surrounding air of a building or vehicle may also contain *soot, smoke, sand, metal powder, and organic or vegetable waste* in the form of *fibers or pappus, fungus spores*, etc.

Cigarette smoke is another air pollutant. It presents a serious problem for passengers with allergies, since it contains compounds causing asthma and other breathing disorders. It also stimulates the development of a number of throat diseases, including *laryngitis* and *tumors*. Dust in the form of aerosols causes lung diseases generally called coniosis, and silicosis is the most common disease affecting people inhaling dust.[†]

Finally, *formaldehyde gases* may be present in the air in new buildings and vehicles. The duration of their release is long (12–13 months) and it creates respiratory problems to hypersensitive inhabitants/passengers in the form of asthmatic and allergic reactions and eye and skin inflammation. The admissible level of formaldehyde concentration in the air is 1 ppm attained in 8 hours or 0.1 ppm attained in a longer time.

Providing an acceptable clean indoor environment is an important engineering task that should be maintained in industrial buildings and vehicles where hazardous-to-health compounds are released. In public buildings and passenger compartments, where visitor/passenger concentration varies between 3 and 6 persons/m^2, removal of the products of human metabolism (carbon dioxide CO_2, water vapor H_2O, methane NH_3, and body odor) is imperative not only because of the resulting decrease of oxygen concentration in the air, but also because of the risk of mass infection and pandemic.

There are four basic methods of air cleaning (purification) and control of the contaminants in rooms and passenger compartments (applied independently or combined with each other)

1. Source elimination or modification
2. Use of outside air to remove contaminants (space ventilation)
3. Local ventilation
4. Air cleaning

The first method is the most effective one when the contaminant can be visually identified—cigarette smoke, mold and fungi, paint, or combustion products. It simply eliminates the contaminant source (no smoking in public transport, offices, and public houses or destruction of mold plantations). Yet, considering other types of contaminants (CO_2, for instance) which have no fixed source or are distributed within the entire building volume, this method is inapplicable.

The second method uses "*space* ventilation," and it is applicable in rooms where the contaminant source occupies the entire volume. The scheme of air exchange ventilation

---

[*] In 1977, an airplane with 54 passengers on board landed in emergency. It stayed for 8 hours on the ground, while the ventilation system was switched off. Three days later, 72% of the passengers had contracted the flu (it was found that one of the passengers was sick prior to the flight).

[†] Embrittlement and calcification of the alveoli.

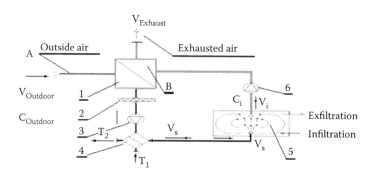

FIGURE 6.8
Space ventilation (without recirculation): 1—recovery system regenerators; 2—filter; 3—supply fan; 4—heat exchanger for thermal treatment of the outside air; 5—object (polluted volume); 6—exhaust fan.

using outside *"fresh"* air to purify the inhabited space is shown in Figure 6.8). It has three regimes of operation

- Total discharge of the polluted air (louver A open and louver B closed)
- Total air recirculation, purifying the polluted air (louver A closed and louver B open)
- Partial discharge of the polluted air (A and B are partially open)

The first regime provides the best air cleaning, but it is nonprofitable from an energetic point of view. Here, the total amount of fresh air is thermally processed (heated or cooled down depend on the current season) before introducing it into the area, and this operation needs extra power.

The second regime is the most profitable one from an energetic point of view, but efficient ventilation depends on the quality of filter 2. Considering that not all contaminants (CO_2 or NO_2, for instance) can be mechanically and effectively removed, we may conclude that the regime has limited application.

The optimal regime providing a reasonable compromise between the energy efficiency and the quality of air cleaning is the third one. An adequate relation between the intake of fresh and recycled air can be found solving the following algebraic system:

$$V_S = V_{Outdoor} + V_{Recir} \qquad (6.2)$$

(volume rate of air supply to the area, m³/s)

$$V_i = V_S + V_{Inf} - V_{Exf}$$
$$V_i = V_{Rec} + V_{Exhoust}$$

(volume rate of air discharge from the area, m³/s)

$$V_S \cdot C_e = V_{Outdoor} \cdot C_e + V_{Recir} \cdot C_i$$
$$V_i \cdot C_i = V_{Recir} \cdot C_i + V_{Exhoust} \cdot C_i$$
$$V_i \cdot C_i = V_S \cdot C_e + V_{Inf} \cdot C_e + V_{Exf} \cdot C_i + N,$$

where

- N, ppm, or g/m^3—contaminant generated in the area
- C_e, C_s, ppm, or g/m^3—pollutant concentration in the outside air and pollutant concentration, currently attained in the area
- C_i, ppm, or g/m^3 is the acceptable pollutant concentration in the area (for carbon dioxide $C_{CO_2} < 1000$ ppm or $1.8\ g/m^3$)

Solving the above system (6.2), we may find the value of the minimal rate of fresh air delivery needed for area cleaning. It should not be less than 6.6 L/s-person (minimal amount of fresh air needed for comfortable habitation).

Practically, one may use specific values of the rate frequency n of air exchanges to solve purely engineering problems

$$V_S = n \times V_0, \tag{6.3}$$

where V_0,[*] m^3/s is the volume of the occupied area and V_S is the prescriptive volume rate of the outside air.[†]

The third method—*"local,"* is ventilation very close to the first one with respect to methodology and capabilities. The basic requirement to its application is the identification of location and type of the contaminant source (for instance, technological machine or instrument generating contaminant).

The *method* is applied in two steps

- *First step*: "isolation" or "encapsulation" of the contaminant source
- *Second step*: local air cleaning and contaminant source neutralization

The first step is realized by fixing the physical volume using fencing walls or by generating sub-pressure around the source area. Source removal and neutralization are performed employing mechanical, chemical, photocatalytic, or membrane technologies.

The fourth method used for *air cleaning* by ventilation is part or continuation of the second and third ones. Air cleaning is technologically accomplished by separating the contaminants from the main air volume. The separation methods depend on

- Size and shape of the pollutant particles
- Particle specific weight
- Particle concentration
- Particle electrical properties
- Particle chemical composition

---

[*] In the Standard 62-1999 next formulas for the supply volume rate of outdoor air calculation are prescribed
$$\dot{V}_{Outdoor} = N_{Occupants} \times \dot{V}_{St}; \dot{V}_{Outdoor} = A_{Occupancy} \times \dot{V}_{St}.$$

[†] The values of the prescriptive delivery rate n or $\dot{V}_{St}$ of the outside air are published in all reference books on ventilation.

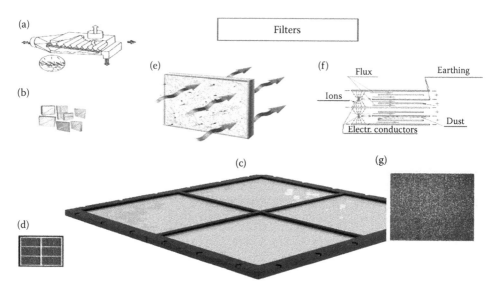

FIGURE 6.9
Filters: (a) mechanical inertial filter; (b) nanofilter; (c) fiber filter; (d) carbon cassette filter; (e) HEPA filter; (f) electro-elastic filter; (g) carbon filter.

The physical devices where separation takes place are called filters, and the following manipulations can be done on the contaminants

- Particle filtration
- Particle direct capture
- Particle centrifuging
- Particle diffusing through semi permeable membranes
- Particle capture by electrostatic traps
- Particle treatment with active ions

Figure 6.9 shows some of the basic filter types: fiber filters, filters fabricated from renewable materials, electronic (ionic) filters, or combination between the first three.

Mechanical filters (a) are used to collect rough mechanical dust. Nanofilters (b) are used to collect fungi, mold and microorganisms. HEPA filters (e) collect the smallest particles, bacteria, spores, fungi, and even viruses. Carbon filters contain active carbon and can neutralize odor and spores. UV filters deactivate aggressive gases and have an anti virus and anti bacteria effect. The choice of the filter type is made after analyzing the pollutant composition and finding the dimensions of the dispersed phase.

6.3.2 System of Indoor Environment Control in Ground Transport

6.3.2.1 Requirements of the Indoor Environment of Ground Transport

During the past century (in a 107-year span after the first officially documented and recognized flight of the Wright brothers in 1903) planes became a means of mass transport, while aviation turned into an industry. Today, ground transport is a preferable travelling method for people whose annual income exceeds $30,000. It saves time and is favored over bus and car transport for distances larger than 100–150 km. Except for its speed, ground transport is safe and provides significant travelling comfort. The safety and health of passengers is the basic priority of the EU and US national administrations. Hence, it is subject to control and regulation.[*]

The systems which control the indoor environment are called environment control systems (ECS) in the regulation documents. This term became popular in aviation. Except for the discussed parameters of internal comfort, other parameters are also controlled in airplanes to guarantee comfort. These are

- Air pressure
- Air purity

During flight at a height of 12,000 m, the temperature of air outside the plane is below −40°C, atmospheric pressure is 2×10^4 Pa, and ozone partial pressure is 0.35×10^4 Pa. The plane indoor environment pressure[†] needs regulation since

- Partial pressure of oxygen in the air should not be lower or higher than a specific limit, in order to guarantee proper function of the human cardiovascular system.
- Ruggedness of the plane structure will be endangered if pressure in the passenger compartment is higher than the specified limit.

The ECS automatically regulates pressure upon change of the flight height. Change of pressure in the passenger compartment takes place at a constant rate (during ascending-at rate 2.8–5.0 kPa/s, while during descending-at rate 1.7 kPa/s). This regulation is performed by pressure control valves incorporated in the plane casing. When flight height drops, pressure in the compartment rises and valves close, keeping the air within the compartment. During ascending pressure in the compartment is regulated by opening the valves. The fresh air is thermally processed by the ECS to meet the following requirements

- Contents of CO—1/20,000 ppm
- Contents of CO_2—up to 3% of the volume (or 0.5%)[‡]
- Contents of O_3—up to 0.25 ppm (as is at sea level), for long stay-up to 0.1 ppm

The ventilation system at 2750 m height should provide 6.6 L/s fresh air per passenger (in emergency, it should provide 0.2 kg/min/passenger in 5 min). The temperature in the passenger compartment should stay within limits $24° \leq T_a \leq 27°C$ ($T_a \approx 4°C$ in the load section). Compartment indoor environment should be taken care of for 30 min prior to the flight (in summer, the plane indoor air is cooled down to 27°C, while in winter it is heated to 21°C). Heating and cooling capacity of the ECS is calculated to stand

- Heat released in the external boundary layer and due to friction
- Heat released by passengers

---

[*] The regulation in the EC is performed by the EJAA (European Joint Aviation Authorities), while in the USA-by FAA (Federal Aviation Administration). Regulation is set forth by specialized norms (JAR/FAR), and it specifies the order of plane certification and operation in the EC and the USA.

[†] ECS regulates pressure in the cockpit and passenger compartment, and it should correspond to the atmospheric pressure at a flight height of 2750 m (8000 ft).

[‡] That norm is specified by the FAA.

- Solar radiation penetrating through the windows
- Heat transfer between passenger compartments and other inhabited areas
- Convection between the equipment and the indoor environment

Note that the air temperature conforms to the current season and to the flight height.

6.3.2.2 Description of ECS Employed in Ground Transport

The ECS scheme is shown in Figure 6.10, where the atmospheric air functions as a working fluid. It is blown by the airplane compressor 1 (by the 8th and 15th compressor stage) and is cooled down in the two air turbines. The cross section of the passenger compartment is shown in Figure 6.10b. It is divided into 2–7 thermal zones controlled by autonomous console thermostats. Cooled air in the air conditioning handler 9, mixed with 50% used air from the passenger compartment, is introduced into the compartment through the central air duct 14 mounted in the plane attic 13. The conditioned air is injected with velocity $w_a < 2.8$ m/s through the regulating nozzles 12 into the passenger seat.

The "used" air, having circulated through the inhabited area, leaves the passenger compartment through exhaust valves mounted in the walls and close to the compartment floor. It is collected in the two sleeves of the air duct 11, and prior to filtering by the HEPA filters 10, it is divided into two equivalents flows. One of them is shot out of the plane and into the atmosphere. The other one is used again—the air is cleaned, mixed with fresh air, and directed to the air conditioning handle 9 for further recirculation in the passenger compartment.

Since the working fluid in air conditioning systems is air, the inverse air compression Brayton cycle operates in airplanes. Figure 6.11 schematically shows four versions of air compression coolers used in airplanes—the scheme supplied is that of a classical air compression cooler. Air bleeding from the turbo-compressor 1 cools down in the heat exchanger 2. Then it passes through the air turbine 3 again reducing its temperature. Air moisture is separated in the separator 6 and then the air is drawn by the fan 9 and is mixed with some amount of the recycled air in recovery section 13 (50% of it is exhausted by the fans 11 from the passenger compartment through HEPA filter 12). The air mix thus prepared is injected into the compartment through the main air duct 14—Figure 6.10b.

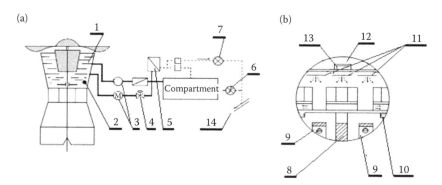

FIGURE 6.10
System for air control: (a) scheme of flow of the "bleed" air; (b) cross section of a passenger compartment: (1) multi-step air compressor; (2) turbine-blade plane motor; (3) air-compression cooler—2 pieces; (4) thermostat; (5) HEPA filters; (6) and (7) hydraulic fuse; (8) air conditioning handler; (9) HEPA filters; (10) duct for used air; (11) regulating nozzles; (12) attic; (13) duct for the conditioned air; (14) relive valve for pressure regulation.

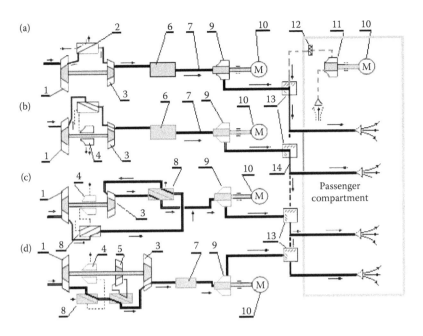

FIGURE 6.11
Schemes of ACM, used in civil aviation: (a) classical scheme; (b) scheme with three rotors; (c) scheme with three rotors and regeneration heat exchanger; (d) scheme with four rotors: 1—compressor; 2—heat exchanger; 3—turbine; 4—fan; 5—turbine; 6—separator; 7—main supply duct; 8—regenerator; 9—fan; 10—motor; 11—return fan; 12—HEPA filters; 13—recovery section.

Figure 6.11b is a modification of the classical air machine. The difference consists in the aggregation of a third impeller (rotor) to the turbo-compressor aggregate (1–3), which is part of the fans for external air 4. It intensifies cooling in the heat exchanger 2 via increase of the delivery of external air, the latter serving as a cold source for the refrigerating machine.

The scheme shown in Figure 6.11c is complicated by replacing the separator 7 with regenerator 8, thus increasing the energy efficiency.

The scheme shown in Figure 6.11d is used in the *Boeing 777*. It is similar to that shown in Figure 6.11b, but it is supplied with additional *regeneration installation* consisting of air turbine 5 and regeneration heat exchanger 8. All passenger planes are designed to have 2 or 3 air-compression cooling machines (ACM). Military planes operate employing open cycles, while most of the civil planes use regeneration schemes.

Control of the ozone concentration in the cockpit and the passenger compartment is strongly recommended; since flight height continuously varies (the ozone concentration is the largest one in April—0.8×10^6 ppm above the sea level). If air with such ozone contents is directly introduced into the compartment, passengers would experience pains in the chest, coughs, choking, fatigue, headaches, nostril blockages, and eye irritation. Oxygen dissociation in planes is performed by

- Air compression in the turbo compressor
- Incorporation of a catalytic ozone converter
- Dissociation in the air conditioning package

Ozone converter dissociates ozone O_3 into oxygen O_2 in a catalytic environment (platinum grid). Its efficiency reaches 95%, and its life cycle is at least 12,000 flight hours.

Another important component of the systems for air environment control is the *HEPA filter* 12—Figure 6.11. It removes the finest admixtures from the recirculating air. These are aerosols, which most often carry bacteria and viruses. The filter cannot be by-passed, since the air returning to the compartment is to be purified. Practice proves that its efficiency increases with the increasing of the term of performance. In planes where smoking isallowed, air quality is controlled by measuring the *concentration* of CO and that of the suspended particles (*RSP*). It is found[*] that the *RSP* concentration varies within limits $40 \leq RSP < 175$ mg/m³.

Another controlled parameter of the plane indoor environment moisture, increases depending on population density and time of passengers' stay on board. The relative air moisture concentration φ_a varies within limits $5\% < \varphi_a \leq 35\%$ (its mean value is between 15% and 20%). If moisture concentration is low (15%–10%), some passengers experience dryness of the eye, nose, and tongue. Yet, there are no indications that this is injurious to health.

Pressure change in the plane indoor environment affects the mechanisms of *oxygen* absorption by the cardiovascular systems of passengers and crew. The amount of oxygen absorbed by blood hemoglobin decreases during plane ascent causing so-called hypoxia. This, together with other effects, causes passengers' fatigue.

Fast pressure change yields discomfort for passengers suffering from respiratory or sinus infections, anemia, or cardiac complaints. Babies and children also feel discomfort during change of the flight height, while adults sometimes experience pain in the middle ear. An additional amount of oxygen is provided during such flight regimes.

6.4 Energy and Light Comfort in Buildings

Quite often, professionals link indoor environmental comfort with only light comfort in the building. Buildings open to the exterior and beautiful nature seen through the window panes attract visitors. Yet, they do not render an account to the reverse of that wonderful sightseeing, that is, the *energy cost* of their delight. The invoice for supplied and consumed energy includes not only hidden expenses for the daily light comfort, but also those for comfort in the dark hours, in the evening owing to artificial illumination. There are also two "dark" sides of the illumination comfort provided by large window panes. The latter diminish the thermal comfort within a large area, since they decrease the indoor radiant temperature T_R of the occupied zone. That effect, in accordance with the bioclimatic chart (Figure 6.1), yields a "cold draught" and feeling cold, although those people may work in bright rooms which offer beautiful views.

Another "secret" of the large window panes is that they add much more kWh-s to the balance of the final energy consumption as compared to *compact* walls. This is so, since heat has a property to find bridges through the building envelope. Envelope glazed components, such as windows panes, window displays, glazed doors, skylights, etc. are classical examples of bridges where heat enters/leaves the building. Before discussing the ability of the envelope's glazed components to aid the transfer of energy, we shall examine problems of light comfort in view of ergonomics.

---

[*] The CO concentration was measured in 94 passenger planes.

Some of the main factors affecting the feeling for indoor environment light comfort are available illumination, its distribution and direction, and the subjective perception of various colors. One may experience light discomfort when

- The external contrast is strong and owing to the brightness of various surfaces within the field of vision.
- Inhabitants are under stress or their eyes are tired.
- Inhabitants are engaged in focusing on small objects or in reading for long periods of time.

When there is light discomfort in a room, people lose their working capacity, experience headaches and total weakness. *Continued* lack of light comfort results in loss of eye contrast sensitivity and eye damage.

The basic light comfort requirement of living area is that the illumination of the working surfaces should be intense enough. Hence, the norms of illumination in different countries and for different people vary indicating the respective *living standard* and local *geography* (sky brightness, vegetation, and landscape colors). Structural design has adopted national illumination standards and brightness parameters. For instance, illumination in day rooms is designed to be 100 lx, while that in libraries 300 lx, etc. That illumination guarantees light comfort for 90% of the inhabitants (only 10% of them feel discomfort). Thus, guaranteeing light comfort is an important engineering architectural task in building design.

The design of comfortable living levels engage three groups of specialists—architects (responsible for building's artistic arrangement), electrical engineers (responsible for building's artificial illumination), and thermal comfort designers (calculating window pane dimensions, thermal characteristics, and the daylight illumination of rooms equipped with windows).

As proved by architectural practice, hybrid lighting systems* are *recommended* for use in South and Middle Europe in view of the moderate local climate. Their supply with energy is twofold: supply with solar energy and supply with energy from other sources (electric or chemical energy, energy of gaseous fuels such as syngas, town gas, propane–butane, or natural gas). Those hybrid lighting systems

- Include a window lighting system (providing facade and upper illumination) mounted within the building periphery or on the roof.
- Provide artificial lighting (most often the lighting fixtures are mounted within the building core). The illumination is switched on in the dark hours.

The synthesis of *such hybrid lighting systems*, conforming to the requirements for minimal initial investments and minimal current overhead expenses for energy and maintenance, is too complicated. This is best illustrated by the operation of systems for automatic control (SAC) of the buildings light comfort (Figure 6.12). In our case we use simulation-based estimations found for a zero energy house (ZEH) in Las Vegas employing the Energy 10 software.

---

* Regarding equatorial, tropical, subtropical and polar climate, public buildings are equipped with artificial illumination.

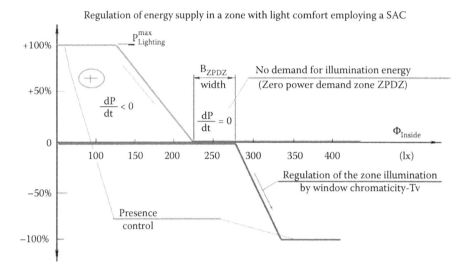

FIGURE 6.12
Regulation of the power of an illumination system which provides light comfort conforming to the demand for energy.

It is seen that there exists a zone for the regulation of switching on and off of the *artificial lighting* where daylight is applied. Hence, no energy for illumination is needed. That zone is called zeroband, deadband, or zero power demand zone (ZPDZ). It is used to avoid accidental switch on due to random fluctuations of the sensor signals (such fluctuations may occur if sensor location is inappropriately chosen, and sensors undergo unplanned heat excursion due to opening of doors and windows, direct radiation, etc.). The zeroband width can be manipulated increased or decreased. The larger the width B_{ZPDZ}, the less the energy consumed by the installation for artificial illumination.

The basic factors affecting B_{ZPDZ} and the illumination of the perimeter zone are the window surface areas A_w (with overall dimensions h and b) and window transparency T_v. The larger the window height h (the surface area A_w, respectively), the deeper the solar light penetration. Moreover, natural illumination will be better and the zeroband will be wider. Hence, that building zone will not require energy for "artificial" illumination for a long period of time, and the overhead expenses for building illumination will be lower.

On the other hand, the use of larger windows, instead of solid walls, will raise the building initial cost, since glazed building envelopes are 2–3 times more expensive than solid ones.[*]

Besides, the use of larger windows which have heat transfer coefficients larger than those of solid walls would result in greater heat leakage/penetration through them. The heating equipment in such buildings is more powerful, more expensive, and consumes more energy to maintain thermal comfort. Thus, an extra parameter is to be considered—thermal energy cost or the relation between the costs of energy for illumination and energy for heating (air conditioning).[†]

---

[*] Luxury walls of polished stones are not considered here.
[†] To solve the optimization problem of calculating the window overall dimensions, one should model the heat exchange between the building and the surroundings using energy-simulating software of the type EnergyPlus and ESP-r.

Despite a better thermal comfort, an ungrounded decrease of window overall dimensions would increase the need for extra energy to provide illumination comfort due to the reduced natural illumination of the building indoor environment. That *ambiguous* effect of the overall dimensions of envelope-glazed components suggests the employment of an optimization procedure to solve the problem of envelope design. Its profit function should be the net present value of the initial investment in the building envelope plus the future payments for energy needed for building heating and illumination.

6.5 Estimation of Needed Energy

Following the strategy of creating a cost-effective insulation of the building envelope, we assume that to besufficiently insulated, a building should satisfy the inequality:

$$U_{Buil} \leq U_{Env}^{Rec}, \tag{6.4}$$

where

- U_{Buil}—assessed value of the total heat transfer coefficient of the designed building envelope (current value)—(see Equation 224 in Dimitroff, 2013)
- U_{Env}^{Rec}—recommended value of the total heat transfer coefficient of the building envelope ($U_{Env}^{Rec} = 4.84(V_{Buil}^{0.666}/A_0R_{Ref})^*IEE_{Env}$ or $U_{Env}^{Rec} = 4.84U_{Ref}(V_{Buil}^{0.666}/A_0)^*IEE_{Env}$)

Here

- IEE_{Env}—building envelope index of energy efficiency[*]
- A_0 and V_{Buil}—area of the surrounding surface and volume of the designed building
- $U_{Ref} = (R_{Ref})^{-1}$—benchmark (reference value) of the overall heat transfer coefficient of the energy standard (subject to periodic normalization in DBA, depending of type of climate—see in Figure 5.17)

In the proposed methodology, the main definitive relationship—Equation 6.4 is used to calculate the recommended value of the total heat transfer coefficient of the building envelope U_{Env}^{Rec}, which is a function of the class (CEE_{Env}) and index (IEE_{Env}) of energy efficiency.

The use of Equation 6.4 to calculate U_{Env}^{Rec} is illustrated in the first row on Table 6.3. The original data used in Table 6.3 concerning our estimations are building volume 567 m³, area of the surrounding surface 450 m², $R_{Ref} = 2.5$ m²K/W (moderate climate) and energy efficiency class of envelope B.

As seen in Table 6.3 the recommended value of the total heat transfer coefficient of the building envelope is $U_{Env}^{Rec} = 0.26$ W/m²K. In the process of the architectural design, the envelope structure should be varied so that the value of U_{Buil} to achieve the necessary and calculated by Equation 6.4 value of the total coefficient of thermal conductivity U_{Env}^{Rec}.

---

[*] See IEE_{Env} in the Correlation table—Table 8, Ap (Dimitrov, 2013).

TABLE 6.3

Examples for the Application of the Suggested Methodology

Variable	Used Formulas or Expressions	Examples
Recommended value U_{Env}^{Ref} of the total thermal transitivity coefficient		$U_{Buil}^{Rec} = 4.84 * \dfrac{567.0^{0.666}}{450.0 * 2.5} * 0.875 = 0.26 \text{ W/m}^2\text{K}$ (for Class B – $IEE_{Env} = 0.875$)
Estimated amount of energy for building heating according to CEE_{Env}	(6.5)	$E_{Buil}^{Heating} = 0.418 * 10^3 * \dfrac{V_{Buil}^{0.666}}{R_{ref}} * a * IEE_{Env} * HDD$ $= 0.418 \dfrac{567.0^{0.666}}{2.5} 1.15 * 2.4 * 10^6 = 31.48 \text{ GJ-year}$ (if HDD = 2400 DD and a = 1.15)
Estimated amount of energy for building cooling (annual) according to CEE_{Env}	(6.6)	$E_{Buil}^{Heating} = 0.418 * \dfrac{V_{Buil}^{0.666}}{R_{ref}} * a * IEE_{Env} * HDD$ $= 0.418 \dfrac{567.0^{0.666}}{2.5} 1.15 * 1.2 * 10^6 = 15.74 \text{ GJ-year}$ (if CDD = 1200 DD and a = 1.15)

Obeying condition (Table 6.3), the energies used in the building for heating and cooling are found via the used in building heating (annually[*]), according to CEE_{Env}:

$$E_{Buil}^{Heating} = aU_{Env}^{Rec}A_0 HDD = 0.418 \times 10^3 \, a \frac{V_{Buil}^{0.666}}{R_{ref}} \, IEE_{Env} HDD, \tag{6.5}$$

Estimation of the energy used in building cooling (annually[†]), according to CEE_{Env}:

$$E_{Buil}^{Cooling} = aU_{Env}^{Rec}A_0 CDD = 0.418 \times 10^3 a \frac{V_{Buil}^{0.666}}{R_{ref}} \, IEE_{Env} CDD, J. \tag{6.6}$$

The coefficient "a" accounts for the impact of the leaks through the building envelope, which is in the range (1.15 < a < 1.3).

6.6 Role of the Energy Manager in Specifying the Targets of Energy Saving

The engagement of the energy manager in the formulation and preparation of targets and plans for energy saving (accounting for energy efficient use by a company, firm, building ownership, or industrial system) is important and fundamental. The national targets of energy saving are assumed as a basis of our further considerations. The present book treats

---

[*] HDD—the annual heating degree days of the building location area.
[†] CDD—the annual cooling degree days of the building location area.

an individual example of how an energy manager of a block of flats in Sofia, Bulgaria, EU, could design a plan for energy saving and formulate the respective indicative tasks.

The "First National Plan for Energy Saving" was approved in 2007, a year before Bulgaria's incorporation in the EU.[*] It concerns the period 2008–2016 and fixes the energy consumption in 2007 amounting to 81×10^9 kWh (69.700 ktoe) as a basic quantity. Bulgaria should save energy amounting to 0.81×10^9 kWh (697 ktoe) annually till 2016 or 1% of the basic quantity. In 2016 the Bulgarian economy should consume 40.9% less *petroleum products*, 26.9% less electricity, 14.6% less coal and wood, and 7.1% less natural gas, as compared to its 2007 consumption. Transport is expected to be the main energy saver—saving 31.7% of energy, followed by industry, households (26.6% each) and public utilities (11.6%). Those values present the indicated national targets.

The national plan for energy saving from 2007 and the accompanying official state documents[†] require all energy dealers and customers[‡] to formulate individual descriptive tasks for energy saving. Such customers are the official (2100 in number) and municipal (3773 in number) buildings which should save 244 and 351 GWh, respectively, by 2016.

According to those *documents*, building managers are obliged to formulate *individual descriptive tasks* for energy saving in buildings[§] under their control. It is also decided that all buildings subjected to energy efficiency audit should save up to 50% of the energy planned for their comfort maintenance. The rest of the official and public buildings should also undergo energy efficiency audits[¶].

6.6.1 Energy Efficiency Audit: Engagements of the Energy Managers

Conforming to the national plan from 2007, called "Strategy 2020," all official and residential buildings (with inhabited areas over 1000 m²) should undergo energy efficiency audits. The audits' aim is to identify the model of energy consumption considering energy type and sources and formulate the individual indicative tasks of energy saving.

The energy audit of an object (a building with its energy system or industrial installation) should solve the following problems:

- Find the structure of energy consumption and predict the behavior of consumers
- Prepare variants of changing the structure of energy consumption considering the "indicative targets" of the branch and choose the most appropriate one considering different economic criteria
- Design a plan and program for energy consumption decrease

The energy audit is performed by certified specialists but after winning a competition organized by the building managers in conformity with the law and regulations of placing a public order. The entire procedure of the energy audit of a building and realization of the

---

[*] Similar "national plan and indicated targets" for energy saving is executed by each of the other members of the EC.

[†] Published in the *Official Gazette*, issue 27 of 10.04.09.

[‡] Two categories; building proprietors conforming to Paragraph 19 of the Energy Efficiency Law and proprietors of industrial systems pursuant to Paragraph 33, Sub-par. 2.

[§] Buildings are divided into two categories

- Audited with respect to energy efficiency (1643 buildings in total).
- Nonaudited with respect to energy efficiency (4228 buildings in total).

[¶] A basis of the energy-indicative task is proposed to be that of sector "public utilities" (11.6%), multiplied by a normalizing coefficient.

engineering project is supervised by the energy managers in their capacity of *investment controllers.*

By decision of the building manager, certified energy-consulting companies should be hired in realizing the plan for energy consumption decrease in buildings or industrial installations. These companies can perform energy audits; prepare energy saving programs, auction paperwork, and organize energy project offers including structural-installation operations and control. Finally, they can regulate energy-converting facilities and prepare building passports and energy certificates. Once collected, data for energy consumption of various users should be systematized and generalized in a so-called "model of energy consumption of the building."

Figure 6.13 shows averaged statistical models of energy consumption of two types of buildings

- Individual residential buildings
- Commercial buildings

If consumers are rated in a descending order, they can be arranged in a table (Table 6.4) shown below Figure 6.13. The first position corresponding to the largest energy consumer is occupied by the system providing thermal comfort (26.3% for residential buildings and 38% for commercial buildings). The next positions are occupied by various consumers depending on the building type and mission.

The second position in the column of residential buildings is occupied by a cooking system (24.1%), while that in the commercial building column—by a lighting system (23.1%). The third position is also occupied by different systems regarding different buildings— in residential buildings this is the "entertainment" system (16.0%), while in commercial buildings it is the "office-communication" system (17.0%)."

As a result of the analysis of the energy consumption models of different buildings, designed by processing of enormous sets of data accumulated during the last 60–70 years, it was found that the strongest factor of energy consumption in a building is the *building mission.* This fact is illustrated in Figure 6.14, where plots of energy consumption in

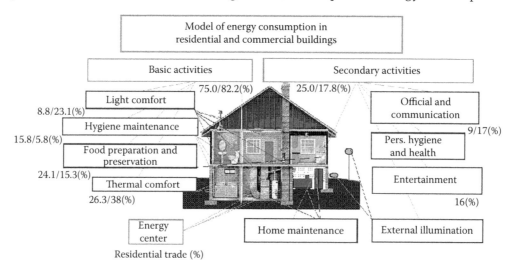

FIGURE 6.13
Structure and rating of the energy consumption in residential and commercial buildings.

TABLE 6.4

Rating of the Energy Consumption in the Residential and Commercial Buildings

System Rating	Rating in a Residential Building	Rating in a Commercial Building
1	Thermal comfort	Thermal comfort
2	Cooking	Lighting system
3	Entertainment system	Office—communication system
4	Washing and ironing	Cooking
5	Office—communication system	Washing machines
6	Lighting system	–

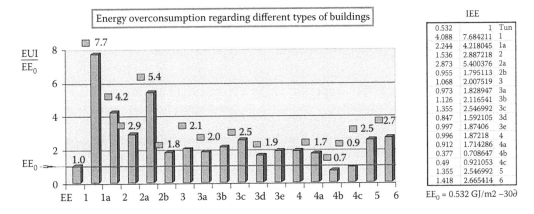

FIGURE 6.14

Estimation of energy overconsumption in public buildings in the period 1990–2002r.: (1) Public buildings (1-lunch room and 1a-groceries); (2) Health service (2-universal, 2a-hospitals, 2b-polyclinics, and consulting rooms); (3) Commercial buildings (3-blocked, 3a-separate, 3b-malls, 3c-warehouses, 3d-public service, and hotels); (4) Social-public buildings (4-offices, 4a-schools, 4b-churches, 4c-law and order, security); (5) General meetings; (6) Other.

different buildings are presented, and data are normalized using the energy standard for commercial buildings[*] $EE_0 = 0.532$ GJ/m²-year. Data in Figure 6.14 prove that the individual indicated tasks of energy managers of residential and commercial buildings should account for building *specificity and activities performed therein*, that is, for the "specificity of the technological processes" popularly speaking. For that purpose it seems useful to divide the different energy consumers into two basic groups

- Main consumers technologically active and conforming to the building mission
- Secondary consumers not directly linked with the building

For instance, the main energy consumers in a residential building (see in Figure 6.15) are those providing thermal and illumination comfort, consuming energy for cooking, food storage, or hygiene maintenance.

---

[*] The standard is approved on the basis of enormous primary information valid for the basic year 2003, and it is supported by *ASHRAE*.

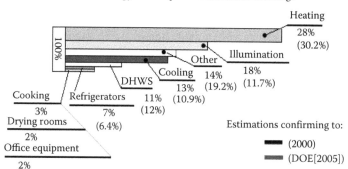

FIGURE 6.15
Model of energy consumption in residential buildings.

Yet, a number of other activities performed in a residential building need energy. These are not basic ones—PC operation, operation of communication, and office equipment, manufacture, etc.

The basic functions of commercial buildings consist in offer commodities and public service. Hence, energy consumed for illumination, office communications, and ventilation should also be added to the basic consumption of energy needed for maintenance of thermal comfort.

The system providing thermal comfort should be designed and applied such as to consume a minimum amount of primary energy,* without violation of the ergonomic requirements. This presumes that building characteristics may generate *"discomfort"* in the shared space, where the users spend much time in selecting articles. To maintain the working capacity of the personnel, however, the basic commercial activities should be performed by authorized personnel in so-called *"thermal oases"* where all conditions of maintaining thermal comfort would be available.

Consumption of energy for cooking and food storage, water heating, etc. should be reduced. Moreover, those activities should be performed in the "cheap energy" time period—0–06 hours A.M.

The energy approach employed in industrial buildings is similar—the energy consumption should be divided into main and secondary aspects. The main consumption is addressed by using more energy-efficient equipment and technology, while the secondary one is shifted to the time span of the cheapest energy.

A plan of energy saving should be developed after specifying the structure of energy consumption and designing its model (separation of the basic consumers from the secondary ones). This should be done based on the previously outlined "indicated individual tasks for energy saving." To fulfill the tasks, the plan should reflect subsequent activities undertaken by the energy manager and approved by the building proprietor or manager. It should also specify terms of realization and contain a list of responsibilities and responsible personnel. Its execution is the basic functional obligation of the energy manager, who should be appropriately trained and authorized.

---

* Conforming to the EU Directive from May 2010, all state and public buildings should have zero energy balance by 2018. That norm should apply to all buildings erected after 2020.

6.6.2 Function of a Building Energy Manager in Attaining Energy Efficiency. Energy Management of Building Installations: A Precondition for Energy Efficiency

The increase of investment activities in the recent 5–10 years was observed mostly on the coast or in large cities. It resulted in erecting a number of hotels, malls, business parks, and large residential buildings. Assume that the process of building design, construction, accommodation, testing, legalization, and commissioning has been successfully concluded. Then, an unexpected problem for investors arises. It consists of how to organize a reliable and economical functioning of the new building, avoiding failure of its systems and installations and efficiently maintaining its operation. Its solution would provide affordable costs of building handling, regular coverage of bank credits, and investment reimbursement. The key to that problem is in the hands of building proprietors who should outline a proper strategy of energy consumption.

To realize that strategy, one should professionally manage buildings, building complexes, and building installations. Hence, an efficient energy-performance structure (traditionally called energy department—ED) should be formed, and properly trained managing personnel should be recruited for efficient building management (Figure 6.16).

The structure of the ED shown in Figure 6.16a is to provide performance of a fully developed energy-supplying system of a building (ESB) comprising all types of installations. In classical public buildings (parliament, ministries, theatres, opera houses, cinemas, etc.) the ED is part of the three-unit general department for maintenance and performance of

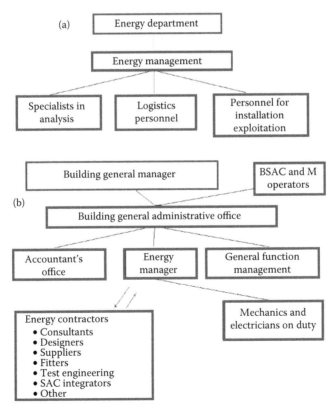

FIGURE 6.16
Structural scheme of the ED of a building: (a) public building; (b) private block of flats with BSAC and M.

the building systems, including units of installation control as well. ED performs monitoring of the energy-supplying system, prepares and analyzes monthly and annual reports on energy consumption, designs the basic model of energy consumption of the particular devices, solar systems, and installations included in the building energy-supplying system. ED specialists design energy saving programs and projects, select and propose projects of energy waste avoidance. Finally, they verify adopted energy saving projects and justify the existence of such a structure of building management. Besides those analytical functions, ED specialists plan preview and manage emergency repairs, control the supply with spare parts, and the availability of fuel, approve repair assemblage, diagnostics, and tests. Finally, they design work schedules of building systems, schedules of dispatchers, operators and technicians on duty, shifts etc., and handle manpower and wages.

This structural scheme is reliable and successfully operates in the public sector. Yet, it is very slow moving and often totally unacceptable for private investors, since it engages a large number of operators and repair personnel (at least 10–14 people). A different version of the structure of the energy management appropriate for private investors is that shown in Figure 6.16b. It is based on ED merging with the building general management department. Its application is possible if the ED integrates with the building general computerized management at BSAC and M level (Building System for Automatic Control and Monitoring). We propose the following synthesis strategy:

- Elimination or minimization of human's participation in monitoring, operative control, and control of data bases
- Setting apart all energy analytical functions and assigning them to independent specialized and licensed companies called energy consultants (see in Figure 6.16b)

Note that the analytical functions include preparation of inspection, standard or detailed energy audits, preparation of energy saving programs and projects, offers of cost-optimized variant of energy leakage avoidance, verification of already adopted energy saving projects via approved tests, etc.

The main figure in that energy management is the energy manager of the building. The manager should be a highly qualified specialist:

- An expert in designing competitive procedures and assigning tasks to contractors (this person should be capable of preparing auction records, negotiating and assessing offers)
- The person should know and participate in all stages of the investment process (and should be experienced in the design of structural engineering systems)
- The manager should be familiar with the operation, structure, and energy-consumption characteristics of the technological equipment (capable of performing a comparative analysis of identical structural systems)
- She/he should know the structure and operation of BSAC and M perfectly, should be familiar with the protocol for exchange of information between different hierarchical levels (possessing practical capabilities of planning the daily, weekly, and annual schedules of operation of the technical installations)
- This person should possess practical capabilities of editing and extracting information from BSAC and M (extraction and analysis of daily, monthly, and annual reports of BSAC and M)

- The manager should possess capabilities of selection, training, and control of the personnel on duty

A popular understanding of local investors and building proprietors is that the problems of energy management arise when the building has been erected, commissioned, and a person has been appointed to handle energy problems. It is quite often assumed that those so-called "energy managers" generate a number of "energy-related" problems for the inhabitants and proprietors. These problems are usually related to discomfort or very high-energy cost payable to energy suppliers or service departments.

The truth is that those problems present the *"birth marks"* of building design, erection, testing, and certification. To be responsible for those activities, people who manage the building and its installations (including the energy managers) should participate in the entire investment process initially—as investment controllers, and subsequently—as managers of building maintenance and performance. They should be competent in designing a detailed, technically proved, and consistent technical–economical project of the building structure and its installations. The technical–economical project of building design and erection (TEPBDE) is an official document that should be taken into account in the preparation of auction paperwork needed to select a building constructor. The preparation of TEPBDE is preceded by a preliminary investigation performed by the proprietor-investor, and its representatives or consultants from all branches of civil engineering. They should take part in all stages of building design and erection and in the starting-synchronization operations, being legal juridical representatives of the proprietor-investor. Finally, they should participate in the check of the structure and its systems, and in the preparation of the structure passport.

Energy managers should be charged with the development and recommendation of successful strategies for reliable and economical building performance, providing favorable conditions of handling bank credits and reimbursement of investment costs.

6.7 Control Questions

1. What does "Indoor environment" comprise?
2. What is the importance of maintaining a comfortable indoor environment for building inhabitants and transport passengers?
3. What groups of engineering technologies are used to provide comfortable habitation conditions?
4. What is the difference between passive and active technologies?
5. Which comfort conditions require the use of energy?
6. What does the condition of thermal comfort consist of?
7. How does human body react to the accumulation of metabolic heat $\Delta Q > 0$?
8. How does human body react if $\Delta Q < 0$?
9. What does heat flux originating from metabolisms depend on?
10. Which are the engineering factors that could affect the heat flux?
11. What does the bioclimatic chart of V. Olgyay present?
12. How can one create thermal comfort under low air temperature T_a?

13. Can one attain thermal comfort under high temperature ($25 < T_a \leq 32°C$)?

14. Which are the criteria for classification of heating (air conditioning) systems?

15. Which are the systems for control of the thermal comfort typically used in residential, public, and commercial buildings?

16. What are the subsystems applied in buildings where the energy demand is evenly distributed within the entire building volume?

17. Which are the advantages of systems with horizontal distribution of the heat carrier?

18. Specify the features of systems creating vehicle thermal comfort.

19. What is the energy model of passenger busses and carriages?

20. What is the difference between the systems for thermal comfort in intercity vehicles and those in city vehicles?

21. Which are the characteristic features of the systems for control of the thermal comfort in passenger compartments of rail carriages?

22. Which are the "waste" products polluting the vehicle indoor environment?

23. Which are the aggressive air contaminants?

24. Specify the most typical neutral pollutant. What is its injurious-to-health effect?

25. Are aerosols injurious to health? Why, if so?

26. Specify the basic methods of air "cleaning"!

27. Which are the basic components of a general-exchange ventilation system?

28. Which is the most energy efficient method of air cleaning?

29. How does the space ventilation operate?

30. Specify the factors affecting the selection of a filter for air cleaning.

31. Specify some types of filters and contaminants that they effectively neutralize.

32. What should be done during technical service of ventilation filters? How does that operation affect energy efficiency?

33. How can one find that the fans belt gear is loose and slips over the belt driving wheels? Does this affect energy efficiency?

34. Who are the passengers preferring ground transport? Why is so?

35. Which is the authorized public body controlling ground transport in the EU?

36. Which are the parameters of the indoor environment that are to be regulated?

37. What is the system of indoor environment control in passenger compartments called?

38. Why is indoor environment pressure in airplanes controlled?

39. What is the reason of performing control of the concentration of ozone (O_3) in the indoor environment?

40. What is the difference between the air temperature T_a in airplane passenger compartments during flight and that in rail carriages?

41. Specify the difference between the thermal load calculated in an airplane and that calculated in a rail vehicle.

42. How does the system for pressure control operate during a plane's descent and at increase of the current pressure (p) in the compartment? How does that system operate during a plane's ascent?

43. Where is the air for the ECS system collected from?

44. How is the ozone (O_3) concentration regulated?

45. How is the cigarette smoke concentration "measured"?

46. What is the effect of air moisture contents? Is low or high moisture concentration dangerous for passengers?

47. What is hypoxia?

48. When does light discomfort of the indoor environment occur?

49. Which are the factors that specify light discomfort?

50. What is the result of personnel working at light discomfort?

51. What is the structure of hybrid illumination systems?

52. How should the natural illumination system be related to the artificial illumination?

53. How does window surface area affect the penetrating energy necessary for the operation of a system for artificial illumination and the energy necessary for the operation of a heating system? How does window surface area affect indoor environment comfort?

54. Knowing that the energy efficiency of lamps with glowing spiral and that of luminescent lamps is low (18%–20%), what is the rest of the portion of converted into?

55. What is the time of operation of the *"First plan for energy efficiency"*?

56. Which year is adopted as a bench mark in assessing energy-saving effects? What is the energy consumption in that year?

57. What are the national indicative tasks of energy saving valid till 2016?

58. What are the national indicative tasks regarding energy type?

59. What are the indicative tasks regarding industrial branches?

60. What is the national indicative task of energy saving in transport?

61. How is an individual indicative task for an investigated building formulated?

62. How is an individual indicative task for a noninvestigated building formulated?

63. What are the methods of analyzing the energy efficiency of a building?

64. Describe the technology of analyzing the energy efficiency of a building.

65. Describe the methods of analyzing the energy efficiency of boilers.

66. Describe the methods of analyzing the energy efficiency of air conditioning installations.

Conclusion

Humanity has passed the stage of development when biomass is burnt for energy. Let us recall what had happened to the forests of Germany and England in the Middle Ages and projections, so that awaits us if instead of wheat and other food crops planted only oil plants that can be converted into bio-diesel. Furthermore, the amount of greenhouse gas emissions will not be reduced if you use the old technology, but on the contrary, used bio-fuel (ethanol, methanol, or biodiesel mixed products) burned in internal combustion engines is not a panacea to the problem because it will create the same volume of greenhouse gases and oil derivatives, but will give the relative energy independence that not have their own oil reserves, inclusive of the EU countries.

Although the buildings and transport itself can become energy efficient, high class (CEE) of equipment and systems, if they are energy systems with technologies whose waste products enhance the greenhouse effect in the atmosphere, their construction and operation is not decreasing but rather increased environmentally. What is the output? The answer lies in the change and development of energy technologies. Until this becomes a reality, the state must establish strict regulations not only for energy but also for eco-efficiency or sustainability. Moreover, the environmental sustainability of the environment in urban conditions must be senior and personal priority. The state administration has the tools to conduct and promote environmental-friendly industrial activities and technology policy, similar to that used in the field of energy efficiency. By analogy with the methodology for assessing the energy efficiency class CEE, which uses an energy efficiency index (IEE) method in environmental sustainability index (IES) is used in the environmental sustainability of the building. This index is a value composed of five different indices (material resources, landscape and urban environment, air, water and soil, indoor comfort and a healthy environment and building energy system—Table C.1) and has an estimated value integral. Detail to determine the index of environmental sustainability of buildings is given in the literature. Environmental sustainability index of the building

TABLE C.1

Correlation Table of the Class Environmental Sustainability of Buildings

Indexes of Environmental Sustainability IES_{Buil} Used in BG_LEED	Class of Environmental Sustainability of the Buildings CES_{Buil} Used in BG_LEED	Comparison between LEED_ US-GBC and BG_LEED
$IES_{Buil} \leq -0.165$	A_0	Diamond
$-0.16 \leq IES_{Buil} \leq 0.59$	A	Platinum
$0.60 \leq IES_{Buil} \leq 1.63$	B	Golg
$1.64 \leq IES_{Buil} \leq 2.54$	C	Silver
$2.55 \leq IES_{Buil} \leq 2.75$	D	LEED™ certified
$2.76 \leq IES_{Buil} \leq 2.8$	E	
$2.85 \leq IES_{Buil} \leq 2.9$	F	
$IES_{Buil} \geq 2.9$	G	

(construction) includes five component indices measuring environmental sustainability with certain weights.[*]

As can be seen from the data given in the third column of Table C.1, similar levels of weight bearing on the final grade is used in the method of US-GBC. The differences are minor and the needs of project evaluations can be ignored.

The introduction of environmental standards and criteria for the construction and operation of buildings and vehicles is a huge challenge for us and the coming generations and humans are the key to the survival of our civilization in the future.

---

[*] See the Mathematical model of the environmental sustainability of buildings in Dimitrov A.V. 2015, *Energy Modeling and Computations in the Building Envelope*, CRC Press.

Appendices

TABLE A.1

Physical Processes and Devices of Energy Conversions

	Mechanical Energy	Thermal Energy	Electrical Energy	Chemical Energy	Radiant Energy
Mechanical energy	Simple machines, gears, multipliers	Friction machines, mixers, heat pumps, chokes	Electro-magnetic inductors, microphones, MHD-generators	Molecular osmosis	Luminescence, cavitation, friction
Thermal energy	Heat engines	Absorber heat pumps and refrigeritors	Effect of Seaback	Endothermic reactions	Incandenset lamps
Electrical energy	Electric motors, electroosmosis, MHD pumps	Effect of Thomson, thermionic diode	Batteries, transformers, PAWES	Electrolysis, electrodialysis	LED, special lamps
Chemical energy	Molecular osmosis	Exothermic reactions, uncontrolled oxidation (burning)	Galvanic elements, fuel cells	Reforming	Chemiluminescence, glowing insects
Radiant energy	Radiometer, Solar rotors	Absorption of light, internal polarization	Photovoltaic cells, internal ionization, barrier layers	Photosynthesis, photodissociation	Fluorescence

TABLE A.2a

Conversion Factors for Energy (Work)

Description	kJ	kWh	kCal$_{15°}$	BTU	m³ gas	Erg
1 kJ	1	2.778×10^{-4}	0.2389	9.478×10^{-4}	2.5×10^{-5}	4.184
1 kWh	3.6×10^3	1	0.86	3411	9.0×10^{-2}	3.59×10^{-14}
1 kCal$_{15°}$	4.186	1.162×10^{-3}	1	3.96	1.05×10^{-4}	0.96×10^{-9}
1 BTU	1055.6	2.931×10^{-4}	0.252	1	2.66×10^{-5}	0.44×10^{-4}
1 m³ gas	39.767	11.05	9500	37.620	1	9.12×10^{-6}
1 Erg	0.239	0.278×10^{-14}	$1.039 \ 10^{-9}$	2.265×10^{-4}	1.096×10^{-7}	1

Note: 1 J = 1 Ws = 4.1868 cal (1 GJ = 109J; 1 TJ = 1012J; 1 PJ = 1015J).
1 toe (tonne oil equivalent) = 7, 4 barrels crude oil—primary energy = 7, 8 barrels—delivered energy = 1270 m³ of NG = 2, 3 metric tons oil.
1 Mtoe (106 toe) = 44.868 PJ (1015J).

TABLE A.2b

Conversion Factors for Power

Description	W	kW	Horsepower	kg f-m/s	Fu-Pound/s	Erg/s
1 W	1	0.001	0.00136	0.102	0.738	1×10^7
1 kW	1000	1	1.36	102	738	1000×10^7
1 Horsepower	736	0.736	1	75	524.5	735.6×10^7
1 kg f-m/s	9.81	0.00981	0.0133	1	7.23	9.81×10^7
1 Fu-pound/s	1.356×10^{-3}	1.356×10^{-3}	1.84×10^{-3}	0.138	1	1.356×10^7
1 Erg/s	1×10^{-13}	1×10^{-10}	1.36×10^{-10}	1.02×10^{-8}	7.38×10^{-8}	1

Note: 1 W = 1 J/s = 1 N m/s.
             1 k.c. = 75 kgm/s = 75 × 9.80665 J/s = 735.5 W.
             1 kcal/h = 4.19 kJ/h = 1.163 W.

TABLE A.3

Physical Properties of Coolant

	Water	R22	R407C	R410C
$T_{кип}$ K ПРИ 10^5 Pa	373.16	232.35	229.6	221.71
r kJ/kg	2250	204.6	201.47	221.04
C_p kJ/kgK	4.186	1.209	1.46	1.519
p_{cr} MPa	22.4	4.99	4.63	4.9
T_{cr} K	647.4	369.29	359.19	344.52
Gas constant R J/kgK	462	96.17	96.47	114.6
Ozone-depleting potential	–	0.05	0	0

Note: R22, R407C, and R410C are used in heat pumps and refrigerators as refrigerants.
             For preheated water vapors C_p = 2.159 kJ/kgK.
             For air C_p = 1.005 kJ/kgK.

TABLE A.4

Computational Formulas for Elementary Thermodynamical Processes

Isochoric $\delta q = du$	Isobaric $\delta q = dh$	Isothermal $\delta q = dl_M$	Adiabatic $\delta l_M = -du$
0	$\delta l_M = p \cdot dv$	$\delta l_M = \dfrac{R \cdot T}{v} dv$	$\delta l_M = -C_v dT$
	$l_{M_{1-2}} = p_1 (v_2 - v_1)$ $= R (T_2 - T_1)$	$l_{M_{1-2}} = R \cdot T_1 \cdot \ln \dfrac{v_2}{v_1}$ $= R \cdot T_1 \cdot \ln \dfrac{p_1}{p_2}$	$l_{M_{1-2}} = C_v \cdot (T_2 - T_1)$
$\delta q = C_v dT$ $q_{1-2} = C_v(T_2 - T_1)$	$\delta q = C_p \cdot dT$ $q_{1-2} = C_p(T_2 - T_1)$ $= h_2 - h_1$	$\delta q = R/v \cdot dv$ $q_{1-2} = R \ln \dfrac{v_2}{v_1}$ $= R \ln \dfrac{v_2}{v_1}$	$\delta q = ds = 0$ $s_{1-2} = const$

TABLE A.5

Mathematical Expressions for Characteristics of the State of the Water and Water Vapors

Phase\ Characteristics	Liquid	Wet Vapor $x = \dfrac{v_x}{v''}$	Dry Vapor
Heat	$q' = C_v\, T_s$	$r = h'' - h'$	$r = u'' - u' + p(v'' - v')$
Internal energy	$u' \approx q' = C_v T_s$	$u_x = h_x - p \cdot v_x$	$u'' = u' + r' + q' + r'$
Enthalpy	$h' = u' \approx q = C_v T_s$	$h_x = h' + r \cdot x$	$h'' = h' + r$
Entropy	$s = 4.19\ln\dfrac{T_s}{273}$	$s_x = s' + \dfrac{r \cdot x}{T_s}$	$s'' = 4.19\ln\dfrac{T_s}{273} + \dfrac{r}{T_s}$

Note: T_S—Temperature of saturation.

TABLE A.6

Energy Capacity of Various Fuels in MJ, koe, or kWh

	Energy Sources	MJ	kg Oe	kWh
1	Coke	25.8	0.676	7.917
2	Hard coal	17.2–30.7	0.41–0.7	4.78–8.53
3	Briquettes (made by brown coal)	20.0	0.48	5.56
4	Lignite	10.5–21.0	0.25–0.5	2.92–5.83
5	Brown coal	5.6–10.5	0.13–0.25	1.55–2.92
6	Oil shale	8.0–9.0	0.19–0.22	2.22–2.5
7	Peat	7.8–13.8	0.19–0.33	2.17–3.83
8	Briquettes made by peat	16.0–16.8	0.38–0.4	4.44–4.67
9	Natural gas	47.2	1.126	13.1
10	Heavy diesel oil	40.0	0.955	11.11
11	Light diesel oil	42.3	1.01	11.75
12	Fuel/gasoline	44.0	1.051	12.222
13	Kerosene	40.0	0.955	11.111
14	Petroleum gas—liquefied	46.0	1.099	12.778
15	Natural gas—liquefied	46.19	1.079	12.553
16	Wood/25% wet	13.8	0.33	3.83
17	Wood pellet/briquettes	16.8	0.401	4.667
18	Garbage	7.4–10.7	0.177	2.06–2.97
19	Produced thermal energy	1.0	0.024	0.278
20	Produced electrical energy	3.6	0.086	1.0

TABLE A.7

Diagram of Physical States of Refrigerants R-134-a and R-404A

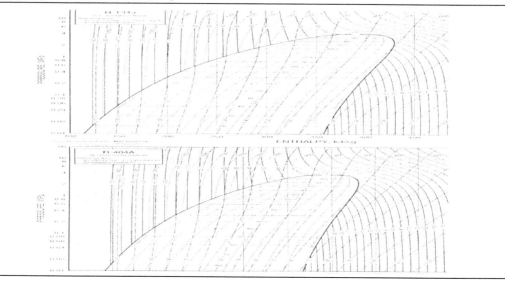

TABLE A.8

Diagram of Physical States of Refrigerants R-410A and R-507A

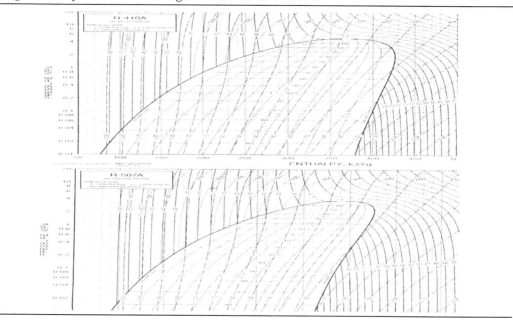

References

References

Cook N., A. Rossi, 2015, *On the Nuclear Mechanisms Underlying the Heat Production by the E-cat*, Cornell University Library, Ithaca, NY, www.arxiv.org, 11 pages, 5 figures, 2 tables, v1&v2.

Dimitrov A., 2013, *Practical Handbook of Finte Element Method for Civil Engineers*, EPU, Pernik, Bulgaria. ISBN 978-954-2983-11-8.

Dimitrov A.V., 2015, *Energy Modeling and Computations in the Building Envelope*, CRC Press, Portland, OR ISBN-13: 978-1-4987-2320-6.

Fleischmann M., S. Pons, 1989, Electrochemically induced nuclear fusion of deuterium, *J. Electroanal Chem.*, 261, 301 in vol. 263.

Hurst D., 2010, Research forecasts accelerating hybrid locomotive sales from 2015–2020, www.green-carcongress.com (published on 31.10.2010).

Parkhomov A., 2014, Исследования аналога высоко температурного теплогенератора Росии, халодный ядреный синтез и шаровая молния, Семинар в РУДН, 25 декабря 2014. Research analogue high temperature heat generator, cold nuclear fusion and fireball, seminar in RSU, December 25, 2014.

Radermacher R., Y. Hwang, 2005, *Vapor Compression Heat Pumps Refrigerant Mixtures*, CRC Press, Boca Raton, FL. ISBN: 10: 0-8493-3489-6.

Rossi A., 2015, Fluid Heater, Patent No. US 9, 115, 913 B1, form 08.26.2015.

Bibliography

ASHRAE Handbook, 2008, *HVAC Systems and Equipment*, SI Ed., Atlanta, GA, ISBN 978-1-933742-34-2.

ASHRAE Handbook, 2009, *Fundamentals*, SI Ed. ASHRAE, Atlanta, GA, ISBN 978-1-933742-55-7.

ASHRAE Handbook, 2011, *HVAC Applications*, SI Ed. ASHRAE. Atlanta, GA, ISBN 978-1-936504-07-7.

Baierlein R., 1999, *Thermal Physics*, Cambrige University Press, UK.

Bauman R., 2002, *Moderm Thermodymanics with Statistical Mechanics*, Clarendon Press, Oxford.

Carrod C., 1995, *Statistical Mechanics and Thermodynamics*, Oxford University Press, Oxford.

Carter A.H., 2001, *Classical and Statistical Thermodynamics*, Prentice-Hall, Upper Saddle River, NY.

Challis L.J., 2003, *Electron-Phonon Interactions into Closed Space Structures*, Oxford University Press, Oxford.

Dimitov A.V., 2008, A new statistical methodology for DELTA_Q method collected data manipulation, *HVAC Res. J.* 14(5), 707–718.

Dimitroff A.V., 2006a, An approach and criteria for heat insulation efficiency estimation, LBL Seminar-Report, March 2006, EETD, Lawrence Berkeley National Laboratory, California.

Dimitroff A.V., 2006b, EnergyPlus and Energy10 simulation results comparison. CER Seminar Report, July 26–29. 06, Las Vegas, Center for Energy Research at UNLV, Nevada.

Dimitroff A.V., 2008, HVAC system simulation overview, Lectures in Climacademy course IV, Maria Curi 7th FP, Pamporovo, May 06 2008.

Dimitroff A.V., 2012a, Energy functions of the building envelope, Education, Science, Innovations, Conference Proceedings of the Second International Conference, European Polytechnical University, Pernik, June 9–10.

Dimitroff A.V., 2012b, *The Building Energy System in of Environmental Sustainable Conditions*. Monograph (in Bulgarian). European Polytechnical University, Pernik.

Dimitroff A.V., 2013, *501 Questions and Multiple Choices in Engineering Installation in Buildings*, European Polytechnical University, Pernik.

Dimitroff A.V., R. Boehm, 2006, Methodology for monitoring system verification by comparing it with software simulation data, Transport 2006, Nov. 12–14, 2006, TU "T.Kableshkov," Sofia.

Dimitroff D., A.V. Dimitroff, 2010, Factors and mechanisms of effective government policy of the control the investment activity process in the field of high efficent energy buildings (parts I and II), International Environmental Conference Save Energy, Save Water, Save the Planet, Sofia, June3–4.

Dimitrov A., 2008, *Energy Efficiency of the Buildings, Their Systems and Installations*. Monograph-part 1, 266 p., Sofia. ISBN 978-954-12-0151-0.

Dimitrov A., 2010, The hybrid locomotives time, *J. Railway Transport*, 7/8.

Fletcher A., 2012, How to prove that the Rossi/Focardi eCat is real, www.Lenr.qumbu.com, updated: SEPT.11.2012.

Incropera F.P., 1974, *Introduction to Molecular Stucture and Thermodynamics*, John Wiley & Sons, NY.

Incropera F.P., D.P. DeWitt, 1985, *Introduction to Heat Transfer*, John Wiley & Sons, NY.

Incropera F.P., D.P. DeWitt, 1996, *2002, Fundamentals of Heat and Mass Transfer*, 3th ed., John Wiley & Sons, NY.

Kondeppud D., 2008, *Introduction to Modern Thermodynamics*, John Wiley & Sons, NY.

McQustion F.C., J.D. Parker, J.D. Spitler, 2005, *Heating, Ventilating, and Air Conditioning*, 6th ed., John Wiley & Sons, Inc, NY.

Stöcker G.N., 1996, *Thermodynamics and Statistical Mechanics*, Springer-Verlag, Berlin.

Stocker W.F., 1993, *Design of Thermal Systems*, 3rd ed., McGraw-Hill, NY.

Stowe K., 1984, *Introduction to Statistical Mechanics and Thermodynamics*, John Wiley & Sons, NY.

Tonchev N., A. Dimitrov, 2009, An Aerogell "intelligent" membrane facade technology development a planed experiment, International Conference on Mechanics and Technology of the Composed Materials. September 22–24, 2009, Varna, Bulgaria, Central Laboratory of Physico-Chemical Mechanics, Bulgarian Academy of Sciences, Acad. G. Bonchev St., Block 1, IV km, 1113 Sofia, Bulgarian Istuute of Polymer Research, University of Neverland, Street 12, City, 1234, Country.

Index

Milton Keynes UK
Ingram Content Group UK Ltd.
UKHW052018071024
449327UK00027B/2325

9 780367 573850